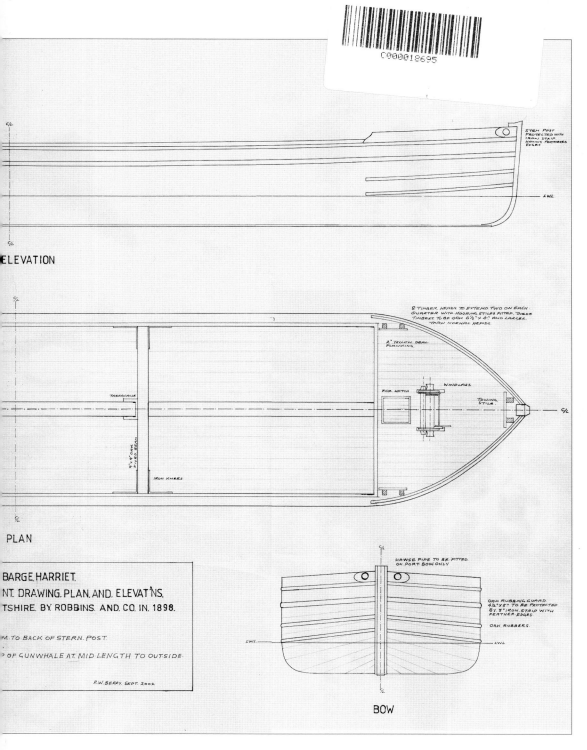

The general arrangement of the Kennet barge Harriet, *prepared as a result of primary research by the author for the building of a plank on frame model, now displayed in the Kennet and Avon Canal Trust Museum at Devizes.*

The

KENNET AND AVON
NAVIGATION

A History

THE
KENNET AND AVON
NAVIGATION

A History

Warren Berry

PHILLIMORE

2009

Published by
PHILLIMORE & CO. LTD
Chichester, West Sussex, England
www.phillimore.co.uk
www.thehistorypress.co.uk

ISBN 978-1-86077-564-2

Printed and bound in Great Britain

CONTENTS

For Katrin and Dillon

LIST OF ILLUSTRATIONS

Frontispiece – Caen Hill flight of locks

Endpapers – The general arrangement (front) and construction drawings (back) of the Kennet barge *Harriet*, prepared as a result of primary research by the author for the building of a plank on frame model, now displayed in the Kennet and Avon Canal Trust Museum at Devizes.

ACKNOWLEDGEMENTS

My thanks go to colleagues and friends at the Kennet and Avon Canal Trust for their help and assistance during the period I was researching and writing this book. I am immensely grateful to Clive Hackford for reading drafts of various chapters, and for advising on many points of detail in the text. My thanks also go to Di Harris for her ongoing help and support, and to Bob Naylor for his assistance in obtaining suitable images and allowing me to benefit from his expertise and knowledge where photographic and associated matters were concerned.

I would like to thank the Trustees of the Kennet and Avon Canal Trust for allowing me to use the Trust's photographic and document archive freely. I am additionally grateful for the assistance provided by staff from the Wiltshire and Berkshire Records Offices, Bath and North East Somerset Archive Office, Reference Libraries at Bath and Trowbridge, and the Curators of the Museum of Bath at Work, and the Wiltshire Heritage Museum at Devizes.

Where illustrations are concerned I am grateful to Colin Green for providing Fig. 65, to the Museum of Bath at Work for Fig. 8, to the Bath Reference Library for Fig. 9, and to Bob Naylor for providing the nine photographs in Chapter 10. The majority of illustrations used in this book, however, were made available from the Canal Trust archive by kind permission of the Trustees. In addition to expressing my thanks to them, I consider it necessary to acknowledge certain individuals who have donated collections and single photographs to the Trust archive and who have requested that they be acknowledged in the event of their publication. Within this group are: Wiltshire County Council Museum Service, Figs 16 and 93; *Waterways World*, Fig. 26; Bath Newspapers, Fig. 29; Jim Lowe, Fig. 35; Michael Ware collection, Figs 40, 74, 92; Harris/Naylor Associates, Figs 34, 42, 45; Derek Pratt, Figs 45, 56-9; Ken Clew collection, Figs 63, 119, 123; Fred Blamplied, Figs 75, 120; J. Priestley, Fig. 79; Philip Wilson collection, Fig. 82; Sue Hopkins collection, Fig. 86; H.D. Astley Hope collection, Fig. 87; Imperial War Museum, Fig. 110; J.E. Manners collection, Fig. 122.

My special thanks goes to my wife, Hazel, for her ongoing encouragement and practical assistance in a number of areas, including numerous textual matters and the organisation and presentation of the images used.

1 | BEFORE THE CANAL:
THE OLD RIVER NAVIGATIONS

In the long period before industrialisation changed the face of Britain forever, the country's roads had undergone little improvement since the time of the Roman conquest, even though various guilds, ecclesiastical associations and the occasional landowner sometimes provided very limited local road repairs. Gradually, road maintenance and improvement became the responsibility of the parishes through which the few main roads that existed were routed. In practice, though, parish officials and the small communities they served rarely had the wherewithal to carry out the necessary work, and local residents had little inclination to travel outside their own immediate areas in any case. This state of affairs was partly due to the insular social arrangements that existed at the time, and partly because journeying by road was often difficult and uncomfortable with highway robbery and assault always a possibility. As a consequence, and because no central body with wider highways responsibilities existed, public roads were badly neglected. Roads became impassable during the winter months, especially for the loaded carts and packhorses that provided the normal mode of transporting goods. Such constraints meant that most economic activity tended to remain local and seasonal in nature, except on the coast, or where the proximity of a navigable river or estuary enabled more viable waterborne trade to take place.

Rivers became an increasingly important means of meeting local transport needs. As the 18th century progressed, the development of industrial processes that consumed large quantities of raw materials stimulated the need for greater improvements in transport arrangements. During this period of development the obvious focus for entrepreneurs, industrialists and engineers was the nation's rivers. It was abundantly evident that, compared with road transport, carrying materials by water was a much more cost-effective alternative, and uniquely suitable for transporting bulk cargoes. More and more rivers were therefore made navigable, although as L.T.C. Rolt has noted, in *Navigable Waterways* (1969), initial river improvements were tentative so as not to conflict with fishing or corn mill interests. Later improvements involved extensive dredging, re-routing of complete river sections, and the removal of fords and flash locks. For the whole of the 17th and much of the 18th centuries, the extent of this work increased substantially.

Schemes for improving rivers, constructing artificial cuts, and for raising the necessary capital by the issue of shares were subject to Acts of Parliament. Often the result of long struggles between conflicting interests, these Acts were important milestones for the promoters of river improvements, but meant bitter disappointment for those who supported the status quo. In the

1   *A train of packhorses and attendants journeying on one of the difficult and dangerous unmade roads that existed at the time.*

period before improvement proposals finally became embedded in Parliamentary Acts, scheme promoters, as well as their opponents, attempted to influence and sometimes coerce the local public and other interested parties. This involved much publicity, statements and counter statements, the collection of petitions, and the provision of evidence before numerous government committees in support of individual cases. Promotional activity of this kind usually took place over an extended period and often incurred considerable costs. When waterway schemes were finally completed, tolls would probably be levied on users of the new facility, so statements of future prospects, engineering surveys, estimates of construction and running costs, together with a statement of the value of capital already subscribed, had to be submitted at the Parliamentary Bill stage. If any of these elements were missing, the Bill was liable to be

thrown out and the promoters would have to start all over again.

Irrespective of all this activity and inevitable conflict, numerous river improvement schemes were put before Parliament during the period. Two such schemes were independently instigated to meet local needs, but the resulting waterways would be eventually joined together by a wide canal section to form the Kennet and Avon Navigation. The proposed schemes were centred on the River Kennet towards the east, and the River Avon in the west of England.

An 18½-mile section of the River Kennet to Newbury, which flowed into the Thames at Reading, was to be made navigable under an Act of 1715. Although trading on the river was common practice, there was much opposition to the new scheme, especially from Reading's mill owners, shopkeepers and other traders. It was feared that Reading's significant commercial advantage, because of its position on the Thames could be lost. An improved Kennet Navigation would result in the town's function as a major trade and distribution centre passing to Newbury.

Despite all the protests, which first started some seven years previously, the Act received Royal Assent, and in September 1715 improvement work commenced. Initial progress was slow, and in *The Kennet and Avon Canal* (1969) Kenneth Clew mentions that the initial lack of engineering expertise was so acute that by 1718 nearly £10,000 had been spent yet only a few new locks had been constructed. It was evident that this unsatisfactory progress could not continue, and later that year, John Hore, the son of one of the Navigation's proprietors, was appointed engineer and surveyor. It is not clear whether Hore had much previous experience of river engineering, although at one time he

had been employed in some survey work on the River Avon. Hore and his labour force soon made an impression, however, and by 1723 they had engineered 11½ miles of new cuts to straighten the river, as well as construct 21 turf-sided locks.

Numerous fixed and swing bridges were built during the period, together with a large dock basin at Newbury that was spacious enough to take 10 large barges at any one time. Great hostility towards the scheme still existed, however, and in 1720 this took a violent turn when an angry mob of 300 men, including the mayor of Reading and other local officials, marched from the town and destroyed parts of the already completed works. The Navigation's proprietors sought to prosecute the officials involved, but agreed not to do so when a pledge was made that no further disruption would occur. Unfortunately, under the influence of strong emotions and bad feelings, this promise

2 *A grain mill on the River Kennet, possibly at Dunmill.*

was soon forgotten, and violent opposition and damage to property continued to such an extent that trading activity on the River Kennet was badly affected for many years to come.

In his book, Rolt also points out that many engineering difficulties were encountered during construction of the Kennet Navigation, amongst which was the steep gradient between Reading and Newbury that amounted to a climb of some 138 feet in a length of just 18½ miles. A number of locks were constructed to overcome this problem, but many of these had to be re-built following severe flooding. These problems, coupled with a general shortage of funds, forced the proprietors to apply for an extension of time to complete the works, and a further two years was subsequently granted

in the Kennet Navigation Act of 1720. Three years later the Navigation was completed, but by 1725, frustrated and unable to convince the proprietors of the validity of his claim for reimbursement of personal funds spent on the Navigation, John Hore had resigned as engineer. Hore continued an association with the Kennet, however, both as wharfinger, or 'wharf manager', at Newbury, and in a maintenance capacity. He was also briefly re-employed to carry out further survey work in 1734.

The Kennet Navigation opened in 1723 at a cost of some £49,000, but the towpath took a further year to complete and is said to have cost a surprising £35,000. Continuing hostility from millers and other local businessmen, together with numerous operational inefficiencies, meant

3   *An artist's impression of Abbey wharf and warehouse at Reading, with a square-sailed barge unloading barrels, c.1700.*

4   *West Mills on the River Kennet, c.1800, showing the swing bridge over the river and weavers' cottages beyond on the right-hand side. West Mills wharf was situated in front of these cottages.*

that trade on the Navigation was less than expected, adversely affecting the company's income from tolls and creating a chronic shortage of funds. The effect of this was low maintenance investment, poor returns for shareholders, and the inability to settle the numerous compensation claims made by landowners and others affected by the waterway's construction. In an attempt to overcome these difficulties, the proprietors, on a number of occasions, explored ways of increasing trade and of reducing overheads. Unfortunately, in practice these actions met with little success. A further Act in 1730 streamlined the powers of the previous two Acts, but whilst this allowed compensation for damage inflicted on the Navigation and its structures, it did little to improve operational effectiveness.

According to Clew, trade on the river mainly consisted of meal, flour and cheese from Newbury towards London, with return cargoes of groceries, coal, timber deals, and heavy goods such as iron. On other waterways during this period, trade in similar commodities meant profitable returns, yet the Kennet proprietors were still unable to make the Navigation pay, even after changes in their organisational structure

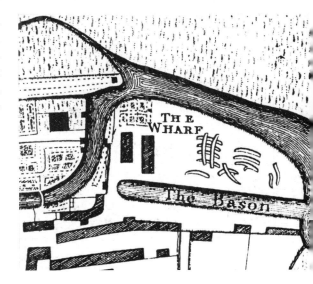

5   *An 1768 map by John Willis showing Newbury wharf and the huge barge basin, or 'bason' as it was then termed.*

allowed individuals with more business-like approaches to become involved.

A local businessman, Francis Page, was soon to alter the Kennet Navigation's fortunes, however. Page owned a successful coal trading operation at Newbury, and in 1760, with a view to gaining a business advantage, attempted to lease the river. Although this offer was refused, the proprietors were impressed by Page's approach and considered him a good and supportive customer. Six years later, Page became a proprietor himself. Clew notes that in a very short time the new proprietor had assumed sole control by buying up the remaining shares. The moment he took control, Page centralised all administration at Newbury, closing the company's London office in the process. He then arranged for outstanding repairs on the Navigation to be completed and paid all current expenses. Numerous other operational changes and improvements were made, some hazardous river shallows removed, the waterway deepened to enable vessels of around four-foot draft to be used on it, and locks also enlarged. These two latter changes allowed access for much larger vessels, the so-called 'Newbury barges' that were capable of carrying 110 tons in their cavernous holds.

Francis Page's management of the Navigation focused on efficiency and economy, and continual improvements, both large and small, soon increased its prosperity. In 1770 a further income was secured when the river was leased to a third party for £1,200 per year, although Page retained all maintenance responsibility. This interim position was altered some years later when Page again resumed full control of the Navigation. He died in 1784 and his two sons took over management of the waterway. By then, however, the possibility of extending the Kennet Navigation by means of a canal was being seriously considered.

At the western end of what was to become the Kennet and Avon Navigation, the River

6 *The 1712 Act for making the River Avon navigable. Legislation by way of an Act of Parliament was required for undertakings such as inland navigations and railways. As the format for all of these was similar, illustrations of the other Acts mentioned in this book have not been included.*

Avon is tidal up to Hanham Mills, some eight miles inland from Bristol, and 18½ miles from where the river flows into the Bristol Channel at Avonmouth. In *The Illustrated History of Canal and River Navigations* (2006), Edward Paget-Tomlinson notes that ships of all sizes had used the busy length of water between Bristol and Avonmouth since the Middle Ages, although this was not the case for higher stretches of the river that lead up to the city of Bath. In fact, as T.S. Willan has suggested in *Bath and the Navigation of the Avon* (1936), Bristol's importance as a major port had probably obscured attempts by the residents of Bath to make the river navigable above Hanham Mills, so that they could also gain access to the sea and take advantage of increased trading opportunities.

The Corporation of Bath had been making tentative plans to improve the river since before the 17th century. In 1700 the Corporation petitioned Parliament for the requisite powers, but opposition from local landowners, farmers and land carriers meant the Bill was soon withdrawn. Some years later a second Bill was promoted, and although those opposing it argued that their livelihoods and ability to pay taxes would be affected by cheap imports from Bristol, Wales and elsewhere, the proposals were supported by Parliament and subsequently enshrined in an Act of 1712.

Unfortunately, work on improving the river did not commence immediately due to existing opposition. B.J. Buchanan has suggested in 'The Avon Navigation and the Inland Port of Bath' (1996) that the Corporation of Bath's status as an elected body may well have made decisive action and constructive negotiations with affected riverside landowners difficult to achieve. Consequently, it wasn't until 1724 that improvement work commenced, and even then it was only because Bristol timber merchant John Hobbs, the Duke of Beaufort, Bath postmaster and local businessman Ralph Allen, and 28 other interested individuals formed a navigation company to which the interests of the Corporation of Bath were transferred. With their associated business interests, the new proprietors all stood to benefit substantially from an expected increase in trading activities generated by a new Navigation. Ralph Allen, amongst others, was involved in a venture to rebuild and expand the increasingly fashionable city of Bath, and expected that the difficulties in transporting goods and building materials by road would be overcome when the Avon Navigation became operational.

John Hore from the Kennet Navigation was appointed engineer, and along an 11½-mile length of the River Avon shallows were dredged, weirs and flash locks by-passed and pound locks built in their place. Thanks mainly to ownership and land dispute issues, however, it was nearly 100 years before a horse towing path was constructed. When drifting or sailing were not practical alternatives, therefore, barges were moved on the Avon Navigation by gangs of men wading the river and towing the laden craft behind them. According to Clew, the cost of making the river navigable from Hanham Mills to Bath was around £12,000, and it was fully opened as far as Bath's city weir at the end of 1727, when the first barge arrived carrying deal boards, pig lead and corn meal.

Clew mentions that in addition to the carriage of goods, passengers were also catered for. Even before the waterway was completed, a passenger boat service using long, light-weight rowing boats known as 'wherries' was operating between the outskirts of Bath and Bristol. Later, when this service had grown in popularity, wherries were provided with enclosed cabin areas for passengers that were proudly advertised as having sash windows.

A steady river trade rapidly developed, and it was only external events affecting the economy, such as the American War of Independence, that caused receipts from tolls to drop temporarily. Timber, stone, metal ores, salt and coal were regularly carried and, with the great days of Bath beginning, foodstuffs, wine, and all manner of extravagent goods were transported on the Navigation. The extent of this trade activity resulted in an expanded waterfront within Bath, and as the requirement for more wharves, quays, warehouses and other associated buildings increased, a busy inland port rapidly developed where the city adjoined the river.

During the period when architect John Wood and others were reconstructing Bath, Ralph Allen was one of the more prominent proprietors of the Navigation. Having retired from his position as city postmaster by this time, the astute businessman was quick to exploit the waterway's potential for his own business interests. Allen had obtained rights to quarry stone from the extensive sandstone beds above Bath, and subsequently obtained the services of John Padmore, a Bristol engineer, to build a railway from the stone quarries to a wharf on the river at Dolmeads, near to where the canal section of the Kennet and Avon would join the river a century later. In *The Pre-History of Railways* (1963), Arthur Elton points out that an advanced type of crane, also designed by Padmore, was located on Dolmead wharf, where it was used to lower blocks of stone onto waiting barges. This 'rat-tailed' crane had a fixed jib, but the whole thing turned on an upright shaft and was fitted with a brake drum and a ratchet and pinion system which prevented the

7   *Although having greater length, passenger wherries were similar in form to this abandoned punt.*

8   *Broad Quay on the River Avon at Bath. At least three wooden cranes, together with a number of horse-drawn carts, are in evidence on the quay. The moored Severn trow, however, is loading or unloading using long wooden planks along which cargoes of barrels, for example, could be conveniently rolled.*

load getting out of hand when the crane was in use.

Padmore's unique railway dropped five hundred feet in one-and-a-half miles, and Elton notes that the rail track itself was constructed of iron-faced timber scantlings on which ran low-loader trucks, guided and braked during their gravity-induced descent by operators precariously perched at the front of each truck. The 13ft long trucks were never turned around and once unloaded were hauled backwards up the hill using two horses and a pulley system. By the use of this arrangement, Ralph Allen was able to reduce significantly his prices so as to ensure a steady trade. The demand for his stone in the rebuilding of Bath was at times so great that two fully loaded stone barges each made four trips a day between Dolmead wharf and various building sites in the city.

Although generally successful, the Avon Navigation was not free from problems and, as Charles Hadfield relates in *The Canals of South and South East England* (1969), some of these problems revolved around the transport of coal. Before the waterway was opened, coal from the Somerset and Gloucester coalfields was transported by road and was used extensively in the surrounding areas. The new Navigation enabled superior coal from Shropshire to be brought down the River Severn and thence up the Avon to Bath. Local miners and road transport workers saw this as a threat to their livelihood, and in 1731 threatened to damage the Navigation. The protest, however, came to nothing, but in 1735 Parliament passed an Act that prescribed the death penalty for wilful damage to the works of a navigable river. Notwithstanding this rather draconian measure, the lock at Saltford was subsequently all but destroyed by a large group of protesters, and a written warning to the Navigation's proprietors stated that unless coal transport on the waterway was stopped, many other such attacks would follow. Although no further damage took place, the proprietors

offered a reward for any information leading to a prosecution. No such information was ever received and the perpetrators were never caught.

Flooding on the River Avon created significant problems for users of the Navigation at certain times of the year, and damage to cargoes, wharves, buildings, boats and barges inevitably resulted. The clean-up after such occurrences was often extremely difficult, particularly as the city sewage and effluent from the many Bath slaughterhouses and breweries all ended up in the River Avon. This state of affairs continued for many years until such time effective flood defence arrangements were eventually developed and put in place.

One unusual maintenance problem encountered by the Navigation's proprietors, as well as by the Company of Merchant Venturers who at the time were responsible for most of the tidal Avon, concerned blocks of stone in the river. When stone barges were being loaded and

9 *The remains of what was probably Dolmeads wharf on the River Avon at Bath. It is not clear when this photograph was taken, but the old wharf was eventually demolished in the early 19th century to make way for the western entrance of the canal section.*

unloaded, along the length of the Navigation and in the tidal section from Hanham Mills down to Avonmouth, pieces of stone and rubble frequently fell into the river. Loading was normally carried out with basic equipment, and a certain amount of such wastage was probably allowed for. An accumulation of stone on the riverbed, however, caused hazards on the Navigation. Ships forced to anchor on an ebbing tide, some three miles west of Bristol at Hungroad, were occasionally damaged when they settled on stone pieces that river and tidal flow had brough downstream and deposited in the moorings. According to a 1785 report, the local haven master, an employee of the Merchant Venturers, found that:

> The quantity of rubble that is brought down from the side banks is the reason for not having that depth of water at Hungroad, formerly there was 17 feet, now but 11 feet at low water.

The proposed solution was the purchase of an especially made set of large iron tongs to remove the stones. The haven master also reported that piles of stone and rubble were being deposited along the riverbank and that many of these were falling or being tipped into the river:

> I took a view of Mr Webber's landing place which is a very bad one being so shelving that many stones lie about half tide down the bank. Mr Hutchens acquainted me that Mr Webber brings large loads of rubble and stuff and stock it on the bank that a great quantity of it must roll into the river and would be detrimental to Hungroad.

The haven master then went on to propose that plank fences should be erected along and below the riverbanks:

> Planks would prevent the said nuisance and would be of no detriment to barges or trows coming down, it would prevent their going aground and would provide a stop to future encroachment on the navigation above the bridge.

He also suggested that a reward be provided for anyone reporting stone tipping activity. It is not evident whether this, or indeed any of the above proposals, were ever implemented in practice.

From 1770 the proprietors sought an extension to trading activities by varying and reducing tolls. This enabled the carriage of low profit commodities such as hay and lime to take place on the river, rather than by other means, whilst generating an appropriate return for the carriers involved. Despite the Navigation's general success, and numerous opportunities to extend trading activities into surrounding areas, the proprietors failed to take much interest in proposals to extend the waterway as far as Chippenham. This reluctance may be understandable in view of the company's monopoly position and the adequate financial returns it was making. In the longer term, however, it meant the Avon Navigation's future as an independent business could not be protected, and it eventually became ripe for takeover once the canal section was completed and the Kennet and Avon Canal Company commenced its trading operation.

2 | CANALS AND CANAL MANIA:
THE GREAT NEW VENTURE

River improvements often required the digging of artificial cuts. W.T. Jackman notes that in many European countries canal building had taken place as far back as Roman times, with the Italians, French and Dutch all subsequently constructing canals, albeit on relatively flat terrain. However, in the early 18th century the idea of an extensive man-made waterway, that passed through towns and villages and mile after mile of open countryside, crossed watersheds and went up and down hills and across deep valleys, was hardly credible. But, as the century progressed and engineers began to develop new surveying and construction techniques, canals began increasingly to appear in the landscape. As Clew points out, once the practicality, utility, and relatively high levels of investment return from these early artificial waterways was proved, an extraordinary period of canal mania swept the country, with numerous canals being built and river navigations extended or incorporated into what was rapidly becoming a network of interlinked, water-based transport routes.

Tentative plans to extend further both the Kennet and the Avon Navigations had been under consideration for a number of years, but because of technical difficulties had never developed into serious schemes. This changed in 1788, however, when a group of influential individuals gathered in the Wiltshire town of Marlborough to discuss a possible westward

extension of the Kennet Navigation. During their debate these men became convinced that such a scheme would only gain widespread support if it were to unite the two existing river Navigations, and believed the necessary surveying and engineering expertise was now available to make the idea a practical proposition. As a result the so-called 'Western Canal' project was born.

The group organised public meetings to ascertain the level of interest in the scheme, and to establish if any financial backing might be available for constructing a canal section between the two rivers. Considerable local support appeared to exist, prompting the group to form themselves into the Western Canal Committee under the chairmanship of Charles Dundas, MP for Berkshire. Dundas lived in the area, took a keen interest in local matters, and for a further 40 years would be involved with the canal. The new committee, which also included Francis Page, the saviour of the Kennet Navigation, within its membership, soon began to consider how best to complete the proposed canal extension. Meetings continued through 1788 to seek the approval of interested parties such as landowners and businessmen, and a pamphlet was published that outlined the trading advantages of the proposed canal.

Three engineers were then appointed by the committee to carry out surveys of two alternative routes. As Edgar Paget-Tomlinson

WESTERN CANAL.

MARLBOROUGH, 29th July, 1788.

AT a Meeting of the Committee appointed to take the Opinion of the Public, refpecting an Extenfion of the Navigation of the Rivers Kennett and Avon, fo as to form a direct Inland Communication between *London* and *Briftol*, and the Weft of *England*, by a Canal from *Newbury* to *Bath*,

CHARLES DUNDAS, Efq. in the Chair,

RESOLVED, That a General Meeting be advertized to be held at the Caftle Inn, at Marlborough, Wilts, on *Tuefday* the 9th Day of *September* next, at Twelve o'Clock, at which Meeting the Landowners, and Parties interefted, are requefted to attend.

RESOLVED alfo, That the following Propofitions be fubmitted to the Opinion of that Meeting, *viz.*

1. That the Juction of the Kennett and Avon Rivers will be of Advantage to the Country.

2. That the under-named Gentlemen be propofed to the General Meeting as a Committee to regulate future Proceedings:

The Marquis of Lanfdown,	Matthew Humphries,	
The Earl of Ailefbury,	Andrew Bayntun,	
Lord Craven,	John Awdry,	
Lord Porchefter,	Paul Methuen,	
Sir Edward Bayntun, Bart.	Paul Cobb Methuen,	
John Archer,	James Montagu,	
William Brummelle,	James Montagu, jun.	
William Pulteney,	Samuel Cam,	Efquires ;
John Walker Heneage,	Jofeph Mortimer,	
Lovelace Bigg,	—— Dickenfon,	
John Pearce,	Efquires ;	Francis Page,
—— James,	John Baverftock,	
Charles Dundas,	John Ward,	
Arthur Jones,	R. H. Gaby,	
Ifaac Pickering,	—— Deane,	
John Hyde,	—— Vanderflighen,	

The Members for the Counties of Wilts, Berks, and Somerfet; the Members for the feveral Boroughs in the faid Counties; the Chief Magiftrates of the faid Boroughs; and fuch other Perfons as fhall be propofed and approved of at the General Meeting.

3. That fuch Committee be directed to employ Mr. Whitworth, or fome other experienced Surveyor, to examine the different Tracts by which the Navigation may be carried, and to make his Report to the faid Committee.

Charles Dundas

10   *The 1788 resolutions from the first meeting of the group of businessmen who would form the committee which would become the Kennet and Avon Canal Company.*

points out, in *Waterways in the Making* (1996), land surveying at this time was difficult, time consuming, and prone to error. In the main this was due to the universal lack of geological information and the fact that proper maps rarely existed. As a consequence, surveyors had no option but to travel on horseback over long distances to examine and record the salient features of likely routes, and make frequent trial excavations to compensate for their lack of knowledge of the terrain.

Despite these problems, preparations rapidly continued. As Clew notes, survey reports were submitted towards the end of 1789, with each engineer favouring the more northerly route via Marlborough, Chippenham, Melksham, Bradford-on-Avon, and thence to Bath. Some concerns still existed, however, particularly with respect to the possible inadequacy of water supply at the summit level, or highest point of the proposed route. As this factor was an extremely important one, the Western Canal Committee decided to have a further, more careful, survey carried out. John Rennie, an engineer who had achieved international recognition from his previous work on bridge building and from his management of the Albion flour mills at London's Blackfriars, was chosen to undertake the task. Although Rennie was undoubtedly an excellent engineer who at the time was in great demand, his previous experience of canals was limited to three surveys, all of which had proved abortive and the canals in question had never been built.

Where the Western Canal was concerned, John Rennie's survey found no lack of water, so towards the end of 1790 the Committee resolved to go ahead with the scheme, appointing Robert Whitworth as scheme engineer and Rennie as consulting engineer. Clew notes that the Committee readily accepted Rennie's estimated construction costs of around £214,000, although resolving that no application for the required Parliamentary Act would be made until at least £75,000 had been subscribed. This cautionary approach was well placed for, in spite of the initial interest, a mere £17,000 was received in subscriptions over the next few months; well below the minimum specified. As a result the Committee agreed to bide their time, believing that things might well change for the better in the not too distant future.

It wasn't long, however, before the growing prosperity of the Midlands and other waterways, which had been authorised and built in earlier years, signalled the fact that canals and canal operations had become highly lucrative. After 1792 an extraordinary period of canal mania swept the country, with numerous plans for new canals being submitted and immense public interest in the subject. The columns in both national and local newspapers were soon filling up with 'today's canal news,' and the advantages or otherwise of particular schemes were hotly debated in newspapers, periodicals and public meeting places. With the benefit of hindsight and survey results, many of these proposals were bound to be impractical, although people were still only too eager to invest their savings in canal shares, and the Western Canal scheme was no exception.

Of course, all this frenzied activity took place in the period before the mid-19th-century Companies Act enabled shares to be bought and sold on the Stock Exchange, and organisations that required canal-building capital had no option but to raise it by local and other subscriptions. In *Canal and River Navigations*, Paget-Tomlinson points out that canal shares were considered to have good long-term potential: shareholders could only lose their investment in the event of the canal company becoming bankrupt. Limited liability was rare at the time, and return on investment could be as much as 50 per cent when a canal was completed and trading on it commenced. Canal shares were considered to have good medium to long-term investment potential and were popular. Once an initial deposit had been made, investors were only required to hand over further sums when a 'call' was made, and this process continued until the full nominal value of each share had been paid. On occasions the Act allowed further calls on each share over and above the nominal value, but this was only done with the agreement of the shareholders concerned. In an attempt to encourage shareholders or proprietors to meet calls, interest on the shares was normally paid out of capital, but where sections of a canal were opened piecemeal payments were often made from revenue accounts.

Clew notes that by the middle of 1793 the initial Western Canal share issue was well subscribed. The Canal Company had grown and absorbed a rival group of Bristol businessmen, and now called itself the Kennet and Avon Canal Committee. Sufficient funds were available to the Kennet and Avon Canal Company for it to confidently instruct John Rennie to carry out more detailed survey work on the proposed line of the renamed Kennet and Avon Canal. Once this third survey was completed, Rennie reported that the initial survey results had in fact been incorrect and it was evident that the proposed northerly route would not provide sufficient water at the summit level. As an alternative he recommended a more southerly route via Hungerford and Devizes, maintaining that this would ensure the necessary water availability, attract lower construction costs, and reduce the original time frame by some 18 months. In order to placate the irate inhabitants of Marlborough, who now stood to lose a potentially important amenity and favourable trade arrangements, Rennie proposed that a short branch canal be built from Marlborough to Hungerford.

Whilst Rennie's reasoning was in all likelihood based on his new and more accurate survey results, James Waylan's 1839 publication, *Chronicles of the Devizes*, suggests that the route changes might well have taken place as a result of political lobbying by two Devizes Members of Parliament, who envisaged increased trading opportunities for their town if the canal were routed through it. Whatever the reasons, the Canal Company approved the revision at their August 1793 general meeting. They also recognised that, although comparative costs for

the revised route were less, Britain's growing financial crisis, brought about by the French Revolution and the ensuing conflicts in Europe, had resulted in steeply rising costs everywhere. Consequently, the estimated cost of completing the new waterway without the Marlborough branch was likely to be around £377,000.

Clew also relates that, much to the consternation and disgust of the inhabitants of Marlborough, the Canal Company decided that owing to increased costs the branch canal should not be built after all. Some compensation was provided, however, in an agreement stating that when the canal became operational a rebate would be given on all goods carried that were destined for Marlborough.

During the surveying process, thought would have been given to the type, or 'gauge', of canal to be built. The basic choice for the Kennet and Avon Canal Company was between a narrow cut that would accommodate 7ft wide narrow boats, and a broader gauge on which 14ft wide barges could also travel. Although narrow gauge canals were cheaper to build, broad, or 'wide', canals were more flexible in use and could theoretically maximise trading potential and thus company profit by accommodating both barges and boats. By the time plans were finalised it was evident the Kennet and Avon was likely to be a broad canal, although in practice it was not until the plans were enshrined in the Act that this important decision was finally made.

As was the case for river navigations, Parliamentary approval was required for canal construction, and this was initially sought through a Parliamentary Bill that, if approved, would become an Act. Those opposing the proposed arrangement were expected to draw up a counter petition and provide evidence for their case. In line with this process the Kennet and Avon Canal Company proprietors petitioned Parliament in favour of the proposed new waterway, providing plans and survey information, a list of subscribers,

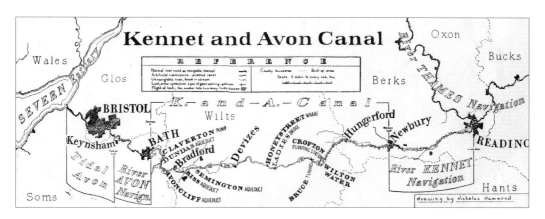

11 *A map showing the River Kennet Navigation from the Thames at Reading to Newbury the canal section from Newbury to Bath and the River Avon Navigation from Bath to Hanham, at which point the river becomes tidal before passing through Bristol and joining the estuarine waters of the River Severn.*

books of reference, and a statement of all associated costs.

At this stage both the Kennet and Avon were still separate navigations, so the Canal Company approached the respective proprietors and both groups immediately offered to sell at a fair valuation. Surprised at this rapid capitulation, the Canal Company decided to take no action with the Kennet for the time being, but to purchase Avon Navigation shares. Later, however, the Canal Company also acquired ownership and complete control of firstly the Avon and then the Kennet Navigations. Discussions also took place with the promoters of other proposed local canals, namely the Wilts and Berks and Somerset Coal Canal, as it was hoped that, when built, these canals would join the Kennet and Avon and thus increase potential markets and extend the trading potential for all concerned. During these discussions it was agreed that once Parliamentary approval had been obtained, construction work on the main canal would continue in such a way that the section from Semington to Bath would be completed by the time the two feeder canals were built, thus allowing trade to start moving and revenue raising to commence.

Once the canal route was agreed and Parliament petitioned, negotiations either commenced or continued with landowners and other parties that had expressed concerns. No powers of compulsory purchase existed in the 18th century, and those affected had to be persuaded to part with land and negotiate subsequent compensation. Opposition to canals came from many quarters: landowners feared the loss of fertile land and water meadows; the canal was considered to be industrial in nature and likely to disfigure estates and bring with it individuals who could well

Disregard game laws and practices, and indulge in acts of vandalism.

Similarly, the operators of grain mills feared a loss of water from the rivers and streams that drove their machinery, and local farmers who needed the mills to grind their corn often supported this stand. Road carriers and traders envisaged a loss of their livelihood, and the owners and masters of small sailing vessels such as sloops, brigs and schooners were concerned that the coasting trade in coal and other bulk commodities would be adversely affected if inland waterways linked the ports to which they also traded. However, as Charles Hadfield has pointed out in *British Canals*, the most sustained opposition came from the towns that were distributing centres for goods moved by road or rivers, who saw themselves being undermined by the construction of canals.

The existence of so many dissenters meant that the process of taking a Bill through Parliament was often stormy and involved bitter argument and debate. Sometimes the dissenters won, but more often the Bill became an Act. In order to achieve such an outcome, offers of compensation, together with the occasional bribe, were often necessary. As well as monetary payments, the compensation awarded to landowners by the Kennet and Avon Canal Company included the construction of both stone and ornamental iron bridges, a suspension bridge, an ornamental lake, and a long, stone-faced, brick-lined tunnel through which the canal passed. The tunnel, the first of its kind on any British canal, was built to hide the waterway, as the landowner in question considered it an offending eyesore and didn't want to see it crossing his land. Passing through what were essentially rural counties, the proposed route of the new canal crossed numerous country estates and passed villages and market towns, and acquiring permissions and negotiating compensation was often a difficult and protracted process. Nevertheless, after extensive preparation and in the face of much petitioning against the Parliamentary Bill, the Act for constructing a canal to join the River Kennet and River Avon finally received Royal Assent in April 1794.

Paget-Tomlinson notes that in much of Europe, and indeed elsewhere in Britain, state funding for projects such as canals was possible. In England, however, capital loans were only made available for financing certain works of a public nature, particularly if these were carried out to alleviate distress linked to the wars with France. As canals were essentially privately constructed local projects, the finance usually had to be raised from local promoters who had a stake in the success of the undertaking. In addition, many years might pass while building continued, during which time there would be little return on money invested. Consequently, the holders of large quantities of shares needed to be relatively affluent and have sufficient funds or income from elsewhere to allow for the medium to longer-term nature of the investment. Where the Kennet and Avon was concerned, subscribers included numerous lay individuals, landowners and gentlemen, as well as six noblemen, twelve clergymen, two municipal corporations, a number of companies, and one university college.

The Parliamentary Act listed 750 shareholders, each of whom was allowed a maximum of 50 shares. It also authorised the raising of £420,000 by means of 3,500 shares of £120 each, and powers were included which enabled the proprietors to raise a further £150,000 from subscribers, or by the use of mortgages. Even so, the Canal Company had to go back to Parliament no less than four times to raise more money, firstly in 1801, again in 1805 and 1809, and finally in 1813. The later share issues were all for values less than those of the original shares.

Now that the main opposition to the canal had been overcome and the legal framework provided by the Act was in place, a general meeting of shareholders was held to elect a committee of management, together with the company officers. Charles Dundas was the obvious choice to chair the 24-strong committee of management, whilst the committee's officers included a clerk and solicitor, a treasurer and accountant, and various engineers and clerks of works, together with their assistants. Three sub-committees were additionally formed so that construction work could be more readily supervised. Each of these sub-committees were meant to control a particular section of the canal route, with the Western District Committee focusing on the section between Bath and Seend, the Wiltshire District Committee on the section from Seend to Wooten Rivers, and the Eastern District between Wooten Rivers and Newbury. This three-district arrangement was later altered, with the Wiltshire District being absorbed into the other two.

In May 1794, according to Clew, John Rennie was appointed engineer for the scheme, a natural progression from his position as consultant to the committee. This appointment meant that Rennie would now have overall responsibility for all engineering, design and construction work. To facilitate control and to ensure effective delegation, Rennie developed a system of working through a number of resident engineers, each of whom was responsible for a particular district. However, the high level of canal-building activity elsewhere in England meant that resident engineers of ability were not easily found. It wasn't until March 1795 that a suitable resident engineer for the important western district was appointed.

Although Rennie was the principal engineer for the Kennet and Avon, the advice of fellow canal engineer, William Jessop, was also occasionally sought, and a short time before work on digging the canal started, Jessop recommended a major change. In essence, as

12 *The Kennet and Avon Canal Company offices in Bath.*

Clew points out, Jessop proposed to alter the canal route so as to avoid the necessity of building a two-and-a-half-mile tunnel near the summit level, and thus make substantial savings. As it stood, additional locks would be required at Crofton, together with a pumping station to raise water to the canal summit, but it was envisaged that current cost estimates could be reduced by around £47,000 by Jessop's proposed changes, and that construction time could be cut by two years. Noting that John Rennie was in full agreement with these proposals, the committee concurred, and plans were altered accordingly.

Once these refinements to the route were agreed, the necessary management and operational structures put in place and the funding sourced, it only remained for the seal of approval of the Company of Proprietors of the Kennet and Avon Navigation to be struck. After some debate on its form, a simple seal design was chosen by a majority vote but, according to Clew, John Ward, the company's principal clerk, voted against the design, considering it 'a very paltry one' and preferring a more complex design that he himself had produced. But the simple design was adopted, and in October 1794 work on cutting the canal section commenced.

Negotiations with and compensation payments to land and property owners still continued, however, and during the construction of the whole canal some 69 individuals were

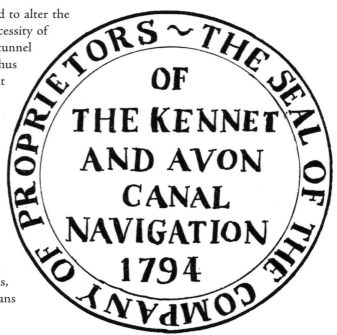

13   *The official seal of the Kennet and Avon Canal Company.*

affected. Compensation was paid in respect of houses and cottages, where buildings were concerned, and gardens, orchards, nurseries, farmyards, fishponds and fields in the case of land holdings. Many of the buildings affected were the homes of tenant farmers and other estate workers, and they and their families had no option but to vacate their homes as canal construction moved forward across the countryside. It is unlikely that compensation would have been given to these unfortunates in quite the same way as that received by those with money and influence, unless of course the individuals concerned had sympathetic landlords who were prepared to provide alternative accommodation.

3 | FORMING THE CUT:
THE BUILDING OF THE CANAL SECTION

The Kennet and Avon Canal was to measure 57 miles, with a climb to the summit level at Crofton, around 450 feet above sea level. Work started at Bradford-on-Avon in October 1794, and progressed towards Bath in the west, as well as in an easterly direction towards Devizes. At around the same time, work commenced at Newbury at the other end of the route and progressed in a westerly direction.

During the 18th century, single contractors rarely built major canals such as the Kennet and Avon because, as Charles Hadfield points out in *The Canal Age*, large civil engineering firms with sufficient capacity did not exist at the time. Instead of this, canal companies either employed their own labour directly, or the company engineer would let a number of contracts for specific sections to locally-based small contractors, or 'hag-masters' as they were called, as canal building progressed. When this was done, work to both excavate and line the canal, as well as build the various structures, was usually let on the basis of so many pence a cubic yard, with deep cuttings attracting extra payments. Once appointed, contractors were expected to provide their own labour, although the canal company often supplied tools and other basic items of equipment.

Hadfield goes on to say that work was usually carried out with a minimum of supervision from the canal company, although ongoing inspection was usually arranged by the canal engineer who utilised the services of 'overlookers' for assessing the quality of work completed, and measuring it so that progress payments could be made to contractors. Direct labour employees were normally paid monthly, or sometimes fortnightly, but could draw advances on a weekly basis if they were in need of more frequent payments. Both direct labour and contractor working were utilised in the building of the Kennet and Avon, although the tendering system used was far from ideal and was often abused. Contractors who had tendered for more than one section of the canal at a time sometimes became overstretched, and their previous experience of major civil engineering projects was insufficient in practice to cope effectively with the required work.

Skilled artisans such as stonemasons, brick-layers and carpenters were employed from the surrounding areas as canal construction progressed. Local engineers and blacksmiths were contracted to provide all ironwork, whilst local hag-masters normally hired and fired the labourers who carried out the actual excavation work. Originally known as 'cutters', and later as 'navigators' or more simply as 'navvies', canal workers were often displaced agricultural labourers, workers experienced in digging and embankment work, ex-miners and quarrymen, and Poor Law vagrants who had little choice but to undertake the heavy and often dangerous work involved in canal excavation and building. These were rough, tough, hard-drinking men, who lived apart from normal society in temporary navvy settlements that grew up alongside the lengthening canal. As the construction gangs moved through the countryside they frequently brought consternation to the inhabitants of remote rural villages that had not experienced such disturbances since Royalist and Parliamentary armies crossed the land in the days of the English Civil War, some 150 years previously. In his book, *Navvyman* (1983), Dick Sullivan describes navvies as working in geographical and social isolation, noting that they were crude, muck-caked men, on a helix spiralling down from prejudice to more isolation, isolation to more prejudice. Eighty per cent were English and most of their work was in England, yet they lived like aliens in their own country, often outside its laws, usually outside the national sense of community.

The wives and families of some navvies opted to live with their men folk in the navvy settlements, but women were on occasion also employed as labourers on excavation and other such work. Although this practice was fairly common on some canal construction sites, it appears to have been less so on the Kennet and Avon. When it did happen it was unusual enough for John Rennie to mention it in his notebook:

> Underhill's contract is proceeding very badly, he is far behind and there is no probability of his making up his leeway. I am therefore at a loss what to advise respecting the mess. Since women were employed in throwing soil into a ditch which had been made under the seat of the bank I am mindful to use them again.

Archimedes screw pumps were sometimes used to drain canal workings. Consisting of

14   *Detail of the stonework of the Dundas aqueduct showing a mason's mark chiselled into it. Each individual stonemason had a different mark that was used to indicate how much stonework he had completed during a given period. In this way the work could be measured and the mason paid accordingly.*

a wooden tube built up in the same way as a barrel, these pumps contained a long iron shaft and attached wooden helix that was normally rotated with the aid of a horse to lift water up the tube. When not in use the pumps were kept submerged in water in order to preserve them. Apart from Archimedes screw pumps and various pulley-based lifting arrangements, however, no sophisticated mechanical equipment was used at this time for canal building. As a consequence, excavation and construction work was arduous and labour intensive, and was made even more difficult by a lack of effective transport and communications arrangements. By the time construction work commenced on the Kennet and Avon, canal building generally followed a methodology that was used for all such construction projects. The following paragraphs, taken from descriptions included in Edward Paget-Tomlinson's books

Waterways in the Making and *Canal and River Navigations*, outline the general process:

Primarily all excavation work depended on muscle power and tools were simple, invariably consisting of pick, spade, crowbar, and simple wooden wheelbarrow. The wheelbarrow was essential, both to remove spoil and for transporting bricks and stone, larger blocks of which were moved on wooden trolleys. It is interesting to note that in the language of the navvy all this spoil, whether timber, rock or earth, was simply known as 'muck'. Both shallow and deep cuttings were made depending on the actual terrain being crossed. Where necessary, tunnels were cut and when this occurred, mining skills were often utilised. In the primitive and dangerous working conditions that existed accidents often happened, particularly where time and cost constraints encouraged navvies to cut corners. Compensation payments for those who were maimed or killed during the process rarely occurred, but when they did, the canal company invariably stipulated that its generosity was to be viewed as being a one-off occurrence that should never be regarded as setting a precedent.

15   *An Archimedes screw pump being used to drain a culvert at Hungerford in 1909. The photograph also shows a set of sheer legs for lifting, two typical navvies' wheelbarrows, and the long-handled earth scoops that were extensively used for excavation work.*

16   *A section through an Archimedes screw pump showing the barrel construction and the way the screw, or helix, is built into the iron and wooden pump spindle.*

Deep cutting work required a high concentration of labour, and involved the moving of large quantities of spoil using a horse worked gin or jenny to haul wheelbarrows full of spoil up wooden planks that were merely laid on the sloping sides of the canal cut, often at angles of anything up to 50 degrees. Each barrow was guided by a man walking behind it, with the connecting rope that led over the gin wheel to the horse, being affixed by a hook close to the barrow wheel and by rings to the two handles. On the narrow, often slippery surface of the angled planks it was inevitable that if barrow or man slipped an accident would happen, and often did. The spoil removed from the canal cut was normally used for filling and banking as work progressed, but good quality topsoil was retained to finish off canal side banks and for spreading on adjoining land.

In order to ensure that water was retained, the canal had to be properly lined. The method used for this important work involved thoroughly mixing local clay and loam with water, before spreading it in layers over the bed of the canal. Known as puddle clay, or simply 'puddle', this mixture was layered about nine or ten inches thick, with each layer being allowed to dry out, or 'mature', for a number of days. It was important to ensure that each layer still retained some wetness when the next layer was applied otherwise proper adhesion between layers would not take place. Each was keyed by spadework to that below, and final consolidation was carried out by workmen 'tamping' the puddle down with their feet, or by paying a local farmer to drive his cattle up and down the newly laid clay bed. Where the ground was particularly porous, the sides of the canal also had to be lined using stepped layers of puddle, the lowest of which was some eighteen inches thick, and the highest nine inches. Sods of earth were built up into walls either side of the cut to prevent the clay from slipping, and when the puddling was completed

for each section water was let in. This latter action ensured the puddle clay remained moist so that cracking and subsequent leaking could not occur. Filling canal sections with water as work progressed had the added advantage of providing accesss to working boats which could then be used to assist in the moving of spoil and building materials.

Early canal builders allowed the canal line to follow natural contours in the land, and although this meant that canals were longer, their construction was in practice often much simpler. By the time planning and construction of the Kennet and Avon commenced, however, engineers were more adventurous and, driven by political, financial and professional concerns, often took more direct and inevitably more difficult routes.

As the Kennet and Avon moved forward, it was soon discovered that in many places beds

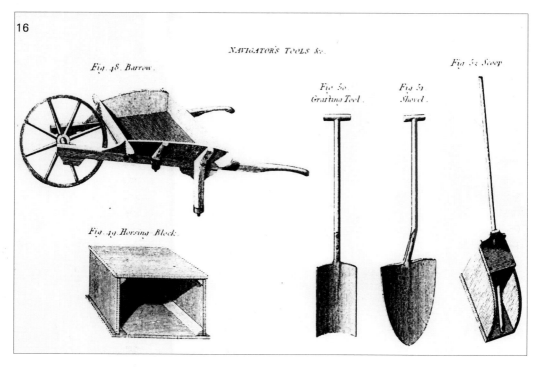

17 *An engraving, dated 1806. This shows the hand tools used by navvies to excavate the canal section of the Kennet and Avon.*

of rock underlaid the shallow clay, a fact that had not been fully appreciated during the initial survey work when it was assumed the route could be rapidly excavated. This discovery was a major setback and meant that on occasion gunpowder had to be used to blast channels through the rock. Blasting was a highly dangerous process and, as D.D. Gladwin points out in *The Canals of Britain* (1973), it was often carried out by merely pummelling the powder and fuse into a hole in the rock before lighting it by hand. Accidents invariably resulted from these primitive practices, and when blasting was taking place a casualty rate of one or two deaths per mile was considered reasonable. Despite these problems, construction progressed at a steady pace and costs were met by regular calls on the share capital available.

18 *Although clearly posed for the photographer, this image shows how wooden planks were used to form roadways so that barrows of spoil, stone and bricks could be more readily moved over the mud and debris surrounding the canal site.*

19 *This photograph of maintenance work on a drained length of canal was taken in 1917, long after the section had been completed. It demonstrates the way in which the canal sides and banks were constructed.*

The main constructional material favoured by the canal proprietors was Bath stone, and large quantities of this were used for building structures such as aqueducts, bridges, locks and wharves. John Rennie considered brick to be a more appropriate constructional material, however, arguing that it was less costly and could be used and transported more readily. Surprisingly, Rennie's advice was not heeded and contractors continued to use stone, either purchasing it from local quarries or sometimes excavating it themselves when a convenient location was found close to the canal line. This latter practice invariably resulted in low quality stone being used.

In *Pre-History of Railways* Arthur Elton mentions that properly selected, quarried, seasoned and laid, Bath stone was, and is, an excellent building material, but inconsistencies in structure and quality mean it can in practice be difficult to use, especially for the inexperienced worker. It is essential to select particular qualities for particular purposes and to avoid stone that weathers badly or is adversely affected by frost. Once selected, the stone must be properly seasoned, which involves stacking in such a way that moisture, or quarry sap, can readily drain away. Unfortunately, these important criteria were not fully considered in the building of the Kennet and Avon, and this resulted in crumbling masonry, defective structures and the requirement for much costly re-working.

According to Clew, by the start of the 1800s the numerous problems being experienced with stone provided by contractors had become so great that the Canal Company decided to augment the dwindling stone supply by opening its own quarry at Conkwell, on the brow of a hill between Bath and Bradford-on-Avon. This quarry was linked to the canal by a steeply inclined double-acting railway track of about seven hundred yards in length, so arranged that when one full wagon descended

an empty one was pulled back up by means of a connecting rope, presumably working around a wheel or pulley system. In practice, the Conkwell stone was little better than that already in use, and as the main Bath quarries at Claverton Down were by this time running out of stone, and in any case charging too high a price for what was left, the Canal Company opened yet another quarry at Winsley. Close to Bradford-on-Avon, it utilised a wooden rail track which led down to the canal at Murhill. Although there was some improvement, the stone from this quarry was unfortunately also below the required quality.

In view of all these difficulties, the Kennet and Avon proprietors' insistence on using stone does not appear to make sense, particularly in light of John Rennie's advice concerning the use of brick as an alternative. The reason for it probably lies in the fact that there was an expectation on behalf of certain influential proprietors that a considerable trade in stone between Bristol and London would arise once the canal was completed, and the goodwill of quarry owners was something that needed to be maintained at all costs.

Clew notes that stone continued to be used as the main constructional material on the western section of the canal between Bath and Devizes, although Rennie was able increasingly to use brick elsewhere. In practice, the quality of brick also caused concern, and problems with availability eventually prompted the Canal Company to establish its own brickworks. For convenience this facility was built close to the canal on a site just outside Devizes. By 1808 better quality stone was provided from Bathampton quarries via the usual inclined railway connection. For a period, quarries at Hanham, between Bath and Bristol, were also used, especially for providing lock facings. The stone was brought up the River Avon by barge.

During the construction process it became evident to Rennie that certain unplanned

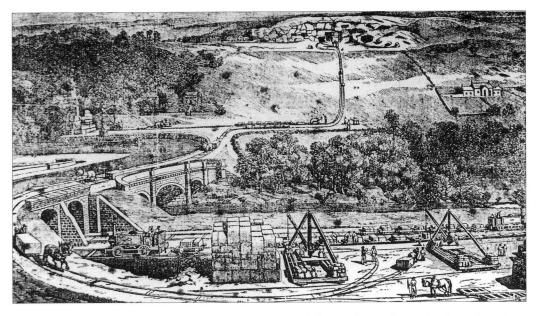

20   *An engraving showing canal construction near Avoncliff. Note the nearly completed aqueduct, the steam engine on tracks, and the horse railway running down from the adjacent quarries. Blocks of stone are being transported and stacked before being cut to size and lifted into place using shear legs and gantries.*

changes to the originally accepted line of the canal would save time and money, and permission for some of these was granted in an Act of 1796, whilst a further Act of 1798 approved the remainder. The 1796 legislation also authorised the construction of a pumping station at Crofton, near the canal summit. As a result of an intervention by the Earl of Ailesbury, who lived nearby, a stipulation was made that the steam engines powering the pumps should consume their own furnace smoke so that the good Earl would not have to see smoke rising from the pumping house smoke stack. An additional stipulation stated that in the event of an associated complaint arising, the member of staff responsible was to be dismissed and never again employed by the Canal Company.

In the same year, a report to the shareholders described the construction then in progress, confidently predicting that work on the 15-mile section east of Bath, which was all out to contract, would be completed within 18 months. It also mentioned that a similar length of canal west of Newbury was now under contract, and that this work would be completed within a 12-month period. Unfortunately, the optimistic tone of the report did not reflect reality, particularly as rising labour and material costs, caused by economic pressures arising from the wars with France, were having an adverse effect. In addition, earlier employment arrangements whereby men had been bound by their work no longer applied, and the higher wages paid by local farmers at harvest time often attracted the canal labourers, who simply walked off the job without giving notice. This practice resulted in large gaps in the labour force that were often difficult to fill. The availability of suitable sub-contractors was also becoming a problem, especially where more specialist work such as that associated with laying foundations for aqueducts and other large structures was concerned.

The Kennet and Avon Canal Company's financial position worsened as costs increased, and Clew suggests that this became even more critical when not all shareholders responded to calls to pay their promised instalment funds. In fact, by the middle of 1796 no less than 822 of the original 3,500 shares were in arrears after six calls. Although action was taken to remedy the situation, and errant shareholders were threatened with the loss of their shares, by early 1797 the position was so bad that the Company's engineers had to slow the rate of work in progress. In addition, the proprietors issued an instruction stating that no new work was to be started until sufficient funding was available in the form of available cash. It was also agreed that work on the middle section of the canal should be stopped until both the western and eastern ends had been completed.

But progress on the eastern end continued and by 1798 the whole 15-mile section had been excavated, with all the locks either built or nearing completion. The western section was still creating problems, however, and whilst a significant part of the excavation work had been completed, many of the locks and other structures, such as bridges, were still unfinished. In line with shareholders' wishes, the Canal Company then made efforts to complete the remaining uncut section between Claverton and Bath so that partial trading on the canal might commence and some return on the money already invested might ensue. Financial problems still affected progress though, and again work had to be curtailed.

Sufficient funds were found, however, to enable construction of a double-tracked iron railway that utilised horse-drawn wagons and joined the completed canal section at Foxhangers to the town of Devizes, pending construction of the Caen Hill locks. This allowed some trade to take place in the area, although barges and boats operating between Bradford-on-Avon and Foxhangers had to be unloaded so that the goods could be transported to Devizes on the horse railway. A similar procedure was in place for cargoes that had reached Devizes from the eastern end of the canal and were destined for locations towards Bradford-on-Avon.

The period leading up to 1800 was hugely problematic for the canal proprietors. According to Clew these problems became even greater during 1800 when financial inconsistencies were discovered in the accounts, and a large amount of clearance and re-construction work was considered necessary thanks to major landslips between Limpley Stoke and Bradford-on-Avon. In March 1800 attempts were made to raise more money by auctioning off shares that had been forfeited through non-payment when calls were made. This action was unsuccessful and as no bids were forthcoming it was decided to merge the shares in question with the rest of the share capital.

By the end of 1800 the financial situation appeared desperate, with little hope of raising more money from existing shareholders whilst the canal remained incomplete. To make matters worse, the Canal Company was also now liable to pay interest on all money advanced. The solution lay in the consolidation of all original shares and the issue of a quantity of new shares, each of which carried a value of £60. At the Annual General Meeting that year, shareholders endorsed this course of action, and it was subsequently embodied in a further Act of 1801.

Apart from a stretch near Bath, the western section of the canal was finally finished in 1801. Not long after, it was decided that the Wiltshire, or central district, grouping should be abolished for administrative purposes and the areas of the eastern and western districts be altered so that they absorbed the Wiltshire district. By the end of 1803 the Canal Company was reporting that the canal was fully open from Great Bedwyn to Newbury in the east, where it joined with the River Kennet Navigation, and that work had started on the Widcombe flight of locks at Bath, the one remaining obstacle on the western section.

The Canal Company report also mentioned that work had commenced on the section from Devizes to Pewsey, although the adjoining section between Pewsey and Great Bedwyn had not been started. In conclusion it was noted that the canal was now open between Bath and Foxhangers, and that the horse railway between there and Devizes was fully operational. Interestingly, John Rennie appears to have had somewhat different views where the latter was concerned, writing in his notebook in early 1803 that:

> The railway from Foxhangers to Devizes is in general badly formed and the rails are badly laid. The sleepers are of wood but too narrow and there is little pieces laid on them to raise the rail. This is in many places split and the rails are loose so that little stability can be expected. The wagons are in general badly formed, some have wheels moving on a short axis like a pulley lying between two cleats of wood, others have axles but the wheels have sharp edges, others again are so large and clumsy they are a load for a horse of themselves. The execution of the railway is very backward.

It is not clear whether Rennie did anything to alleviate these problems, but towards the end of 1804 he reported that a further sum approaching £142,000 would be required to complete the canal. Amidst the uproar this statement created, Charles Dundas attempted to placate the shareholders by promising that a special investigation would be carried out into the reasons costs had again increased. As Clew notes, the investigation was duly initiated, and the report to the Canal Company investigators highlighted frost damage and other problems

that had resulted from the use of defective stone. It also stressed that a general increase in both labour and materials costs had taken place, and pointed out that the canal bed had leaked in a number of places and had to be re-lined. Leakage proved an ongoing problem for certain sections of the canal, although it was recognised that this was due to fissures and faults in the underlying rock strata generating sufficient pressure to cause holes to be blown in the puddle lining the canal. It was also reported that a number of debtors had not paid and that several additional corn mills had to be purchased in order to avoid annual damage payments. The investigators concluded that John Rennie was not to blame for any of these problems, which were all considered to be outside his immediate control.

The issue of how to acquire additional funds still remained, so a further application was made to Parliament with a view to increasing the share capital. The new Act received Royal Assent in June 1805 and granted powers to raise another £200,000 to complete the canal. The new share values were initially set at £30 each, but as the new issue was not subscribed to as quickly as expected, values were subsequently reduced to £20 per share. This had the desired effect and the necessary funds were pledged.

Although repairs to several parts of the canal were undertaken in the interim, by June 1806 work had commenced on the remaining section from Pewsey to Great Bedwyn. Quite rapid progress was made and, as a consequence, a report of that year noted that most of the work was out to contract. Although the horse railway between Foxhangers and Devizes was still in operation, work on building the Caen Hill locks had already commenced. A shortage of bricks (two million of which had been diverted to build the tunnel at Savernake) had initially slowed down the work. Construction gradually continued over the next three years and it was believed the canal would be completed by

21 *The 1803 Canal Company management committee report to the shareholders indicating construction progress and associated costs.*

KENNET and AVON CANAL.

THE General COMMITTEE of the KENNET and AVON CANAL COMPANY, acquitting themselves of a Duty Incumbent upon them, and desirous of meeting the Wishes of the Proprietors of that Undertaking; submit to them the following REPORT, as the general State of the Accounts and Works,

From Bath to Foxhanger, a length of 19 Miles, the Navigation is uninterrupted, and the Junction with the River Avon, by means of Seven Locks, is now in great forwardness;

From Foxhanger to Devizes, a length of 3 Miles, a Railroad is finished;

From Devizes to Pewsey, a length of twelve Miles, the Cutting and Masonry, are in great forwardness, and the Land for the Use of the Canal purchased;

From Pewsey to Great Bedwin, a length of 8 Miles, is not entered upon;

From Great Bedwin to Newbury, a length of 15 Miles, is Navigable.

TO such of the Proprietors who have never visited the Works, or viewed the Line of the Canal, the Plan annexed hereto will, it is hoped, be particularly acceptable as well as useful: This will also shew the Places at which the Coal Canals join the Kennet and Avon (at Monkton Comb) and where the Wilts and Berks Canal communicates with it at Semington.

SINCE the Calls made on the Subscribers to the new Shares, (the two first of which were expended in the Payment of the Interest, due to the Proprietors on the old Shares) the Committee have expended as per Statement annexed.

BATH,
September 23. 1803.

CHARLES DUNDAS, Chairman.

Dr.   The Kennet and Avon Canal Company.   Cr.

1800 Feb.	£	s.	d.	1800 Feb.	£	s.	d
To Amount of Cash received from Subscribers	360,575	2	3	By amount of Expenditure	353,921	6	0
Balance in the Eastern and Western Districts	809	13	8	Building Locks, Bridges, &c.	815	5	6
To Cash received of F. Page amount of his debt	7,985	13	4	Cost of Land, River Shares, &c	2,274	1	8
Do. of Do. Interest	1,753	14	2	Engineers, Surveyors, Clerks, Agents, Law Charges &c.	6,466	16	0
Do. Tonnage, on Eastern and Western Districts	2,231	7	8	Earth Work	18,366	18	6
Boats sold	525	0	0	Boats, Craft, Timber and other Materials	16,747	10	11
Peat	88	13	1	Committee Expences	1,116	3	8
Rent of Land &c.	375	19	2	Printing, Stationary &c.	516	5	5
Hay and Grass	91	8	6	Interest paid Proprietors	34,636	0	10
Dividends on Bath River shares	2,092	10	0	Steam and Horse Engines	1,521	11	4
Balance due to John Thomas	81	10	0	Damage paid Land Owners repairing Slips &c.	16,345	2	7
Do. John Ward	235	19	8	Work Middle District	29,522	8	8
Do. Harfords Davis and Co. the Treasurers	4,209	14	0	Cash in hand Western District	242	16	5
Cash received from the Subscribers	101,986	11	4	Do. Eastern District	100	9	3
	£482,492	16	10		£482,492	16	10

the end of 1809. Money was still a problem, however, and a further Act in June 1809 gave powers to raise an additional £80,000 by the issue of £24 shares.

The 1810 shareholders' report announced that the canal had been completed apart from the Caen Hill locks, although these were also opened to traffic by the end of the year. No great celebrations occurred on completion of the canal, as its cost was already far in excess of the original £330,000 estimate, having risen to £950,000. As Clew points out, however, one small celebratory gesture was made at the committee meeting in December 1810, when it was agreed:

That the Treasurers Clerks should be presented with a Christmas present of Five Guineas.

Most canals built during this period, when accurate information on geological and other features was difficult to acquire, invariably cost much more than their original estimates. This was a normal fact of life for canal engineers and shareholders alike. In *Canal Age* (1968), Charles Hadfield indicated that the cost on completion of the Kennet and Avon equated to a staggering average figure of more than £16,500 per mile; one of the largest unit costs of any broad canal built in Britain during the period.

The COMMITTEE of MANAGEMENT of the KENNETT and AVON CANAL COMPANY, (referring to the detailed Account of the COMPANY's AFFAIRS to the 30th November, 1804, in their laſt Report) have now to communicate to the PROPRIETORS the PROGRESS of the WORKS, and a STATEMENT of their ACCOUNTS and EXPENDITURE, from that Period to the 30th ult.

AT the General Meeting of Proprietors held at Bath the 18th inſtant, Two Reports from their Chief Engineer and Superintendent were read, by which it appears that the Works between Devizes and Pewſey, a diſtance of Twelve Miles, are in regular Progreſs, that Four Miles thereof have been completed, and filled with Water *riſing in the Canal*; and although not ſtated in thoſe Reports, it may be ſatisfactory to the Proprietors to be informed, that Four Miles more will be Navigable in March, and the remainder in June.

In their laſt Report the Superintendent's Statements were made up "partly from Eſtimates and partly from Expenditure," the Sum eſtimated to finiſh the Twelve Miles above-mentioned was £. s. d. 39,085 19 9

The Sum ſince expended is - - - - - - - - - - 19,779 4 6

Leaving a Balance to finiſh of - - - - - - 19,306 15 3

The Sum now eſtimated to finiſh is - - - - - - £ 19,400 6 5
From which deducting the above Balance - - - - - 19,306 15 3

There will be an Exceſs of only - - - - - - - £ 93 11 2

The Committee are much gratified in communicating this Information, as it furniſhes an additional Proof, that the Sum ſtated to be neceſſary to finiſh the Canal in their laſt Report, will be ſufficient for the purpoſe, not altogether from the Reports of the Engineer and Superintendant, but from the very near Agreement between the Eſtimates and the Work done, *all the Calculations being made from actual Admeaſurement*, and they are decidedly of Opinion, that no Circumſtance whatever, except an Advance in the Price of Materials and Labour, will cauſe an Exceſs; the Prices for the Works, between Pewſey and Great Bedwin, being eſtimated at a higher Rate than thoſe now under Execution.

Since their laſt Report, ſeveral Parts of the Canal have been under Repair, but notwithſtanding this Circumſtance, the Navigation from Bath to Foxhanger, near Devizes, and from Great Bedwin to Newbury, has not been impeded; the Committee, however, think it proper to ſtate, that during an Interval of Six Weeks, very little Tonnage was received upon the Weſtern End of the Line, in conſequence of ſome Repairs and Alterations upon the Coal Canal, but this Company having now effected a Junction between their Upper and Lower Level upon the Paulton Line, by means of Locks, will be enabled to furniſh a much larger Supply of Coal, and a conſiderable Increaſe in Tonnage may therefore be expected.

The Committee having been informed, that ſeveral reſpectable Proprietors have expreſſed ſome Doubts as to the Probability of an ample Supply of Water, they think themſelves warranted in ſtating, from actual Surveys which have been made and Information they have obtained, that when the Works at the Summit Level are finiſhed, the Supply of

Water will be abundant, and fully ſufficient for all the Purpoſes of the Undertaking; and, in Anſwer to a Variety of Reports reſpecting the State of the Works already finiſhed, the Committee can aſſure the Proprietors, that the General State of the Canal is conſiderably improved, and that the Sums neceſſary for Repairs will not in general exceed what may be expected in a Work of ſuch Extent and Magnitude.

The 10 per Cent. eſtimated as a ſufficient Cover for Superviſals, &c. &c. has been this Year exceeded; but if the Expence of procuring the laſt Act of Parliament and ſome other Articles, which will not again occur, be deducted, it will be conſiderably leſs.

The Difference in Account between the Company and their Treaſurers, Meſſrs. HARFORDS, DAVIS & Co. has been adjuſted compleatly to the Satisfaction of the Committee.

They have only to add, that agreeably to the Reſolutions of the General Meeting, held at Bath the 16th Auguſt, and in conformity to the Act paſſed in the laſt Seſſion of Parliament, for raiſing Money to compleat the Canal and Works, the New Shares were offered to the Proprietors and alſo to the Public, at £30. and Optional Notes at £50. and the Subſcriptions not having amounted to the Sum preſcribed by the Act, the Proprietors at their General Meeting, held at Bath the 18th inſtant, unanimouſly reſolved to offer the New Shares at the low Price of £20. and Optional Notes at £33. 6s. 8d. to *Proprietors only*, under a full Conviction that thoſe Terms muſt operate as an Inducement to fill the Subſcription.

The Committee conſider it their Duty to appriſe thoſe Proprietors who do not chuſe to cover their Shares, that they have a Power, by the Act of Parliament, to transfer their Right of Subſcription by the Appointment of a Nominee or Nominees.

C. DUNDAS, CHAIRMAN.

DECEMBER 31, 1805.

REPORT

OF THE

Committee of Management

TO THE

COMPANY OF PROPRIETORS

OF THE

KENNET AND AVON CANAL.

THE Committee of Management have the satisfaction to acquaint the Proprietors, that the Confidence with which they expressed their Opinion in their Report of July 17, 1810, that the Canal would be completed in the last Year has been fully Justified by the Result ; the whole Line from Newbury to Bath having been opened and navigable on the 28th of December, 1810.

Your Committee deferred making their Annual Report until they had inspected the Canal. In the course of the present Week they have surveyed the whole Line from Newbury to Bath. The Observations they were enabled to make were highly gratifying. The Masonry is deserving of particular Commendation, and the Work in general reflects great Credit on those who have superintended the Execution of it.

It affords your Committee much Satisfaction to be enabled to confirm their former Opinion as to the Supply of Water on the whole of the Line, which they are convinced will be fully equal to any Quantity of Tonnage that may be expected to pass upon it.

Your Committee have the most flattering Assurances of an extended and increasing Trade from various Quarters, and from Persons best informed on the Subject, and they consider the present as a favourable Opportunity for Persons disposed to embark in the Carrying Business on this Canal, as the Number of Boats now employed is not sufficient even for the present Trade; and they hazard little in asserting that when the Public are satisfied by Experience of the Cheapness and Dispatch by this Conveyance, a very great increase of Tonnage may be expected.

The annexed Account is, as usual, a Statement of the Receipts and Expenditure for the last Year, ending 30 May, since which the following Tonnage Accounts have been received, which it may be satisfactory to state to the Proprietors.

In the Month of June	——	1235	12 10
——————July	——	1256	18 8
——————August	——	1878	19 10
Total £		4371	11 4

In the last Session of Parliament, Acts were passed for making a Canal from the Kennet and Avon at Bath to Bristol, and for making a Canal to be called *The Bristol and Taunton Canal*, from the Vicinity of Bristol to join the Grand Western Canal at Taunton.

The Committee regard both these Canals as highly important, and beneficial to the Kennet and Avon. By the former the Navigation from Bath to Bristol will be relieved from the Impediments to which River Navigations are always liable, and an easy Access may be formed with the Collieries in Glocestershire; and by the latter the whole of the West of England will be opened to the Kennet and Avon. Great as these Advantages may be, the Committee look with no less Satisfaction to a Communication with the Nailsea Collieries, by means of a collateral Cut, for which Powers have been given by the last mentioned Act.

Since this Act was passed two Engineers of Eminence have been employed to examine the Country about Nailsea ; they have each made a Report, and their Opinions (formed without any previous Communication with each other) are most decisive as to the abundant Quantity of Coal which may be obtained at a very easy Expence.

The Distance by the proposed Canal from Nailsea will not exceed ten Miles from Bristol, and the Country is so level, that no Doubt is entertained, but the Communication may be completely effected within two Years.

Your Committee are informed, that from this Neigbourhood an immense Quantity of Coal will be furnished to the Kennet and Avon, at a much cheaper Rate than from any other Source, which they have at present the means of bringing on their Canal.

It might be expected that your Committee would at this Time announce to the Proprietors when they might look for a Dividend; but although the Canal was opened in December last, a considerable Sum has been since expended in Locks, Ponds, Towing Paths, Fences, Steam and Water Engines, &c. &c. the Amount of which has not been ascertained, as several Parts of these Works are not yet completely finished; they can however assure the Proprietors, that upon the most exact Calculation it is in their Power to form, the Lands and other Property belonging to the Company will fully cover the whole of this Expenditure.

C. DUNDAS,
Chairman.

Bath, September 27, 1811.

23 *The 1811 Canal Company management committee report informing shareholders that the canal section was fully open and that trading was continuing along the full length of the Navigation. As numerous ancillary structures and facilities were still not completed, dividends could not as yet be paid.*

22 *(Opposite) The 1805 Canal Company management committee report, this time of a more detailed sort and indicating amongst other matters those parts of the Navigation on which trading had already been allowed but also noting that repairs to previously constructed parts had been necessary.*

The Average Prices of Navigable Canal Shares and other
Property, in April 1813 (to the 24th), at the Office
of Mr. SCOTT, 28, New Bridge Street, London.

Monmouth, £111
Grand Junction, £225, £223
Old Union, £101
Grand Union, £27 Discount
Worcester and Birmingham, £30
Ellesmere, £64
Kennet and Avon, £22. 10s.
Wilts and Berks, £18, £21
Huddersfield, £17. 10s.
Regent's, £12 Discount
Ripon, £70
Chelmer, £85
Ashby, £17. £16
Bolton and Bury, £93
West-India Dock, £148
London Dock Stock, £101
Globe Assurance, £105
Albion Assurance, £46
Grand Junction Water-Works, £21 to £21. 10s.
Scotch Mines Stock, dividing £5 per Cent. £105
Strand Bridge, £46 Discount
Vauxhall Ditto, £53 Discount
London Institution, £45
Surrey Ditto, £14. 14s.

The MONTHLY SALE is on the First FRIDAY.

24　*The 1813 document comparing the
average price of shares associated with
Navigations and other related undertakings.
Share values were presented in this way for
investors to make purchases at monthly sales.*

25　*An early photograph
of the Caen Hill flight of
locks, with side ponds and
a towpath in evidence. As
one of the last structures
to be completed, the flight
eventually took the place of a
horse railway that stretched
up the hill to Devizes.*

26 *An unusual view of the canal, flanked on one side by tall poplar trees and with what appears to be a narrow boat or barge in the far distance.*

27 *Caen Hill's dramatic flight of 16 locks, c.1900. No trading activity is in evidence, although a small group of people can be seen standing on the towpath.*

28 *A photograph, c.1900, showing the Kennet and Avon at its junction with the Wilts and Berks Canal near Semington. Known to Wiltshire and Berkshire boatmen as the 'Ippey cut', this junction was very busy from about 1810 to 1876, mainly as a result of coal traffic.*

29 *A photograph, c.1950, showing the bridge which crossed the A36 near Bath being demolished. It once carried a tramway used to transport stone from a nearby quarry down to the canal construction site that was situated below the road on the right-hand side.*

4 | Fashioned from Wood and Iron, Brick and Stone:
The Working Waterway

To enable a canal to function properly as a trading route, various additional structures and features had to be incorporated. The canal-side banks required protection against erosion created by the wash from canal craft, as well as from damage caused when collisions between boats and the bank invariably occurred within the limited confines of the canal. The usual way of achieving such protection was to provide a ledge, or 'berm', just below the water surface on which rushes could be planted to protect and bind the sides and break up the wash from passing craft. Where this arrangement was not considered suitable, other methods, such as stakes and brushwood or wooden pilings, were utilised. 'Winding' holes were another feature of canals. Significantly wider than the rest of the canal, these short sections, often close to wharves, were carefully sited so that passing canal craft would be in sight of each other, and were provided to give canal boats a greater width of water in which they could 'wind', or turn around.

A towing path existed along the whole length of the Kennet and Avon Canal, although it wasn't until 1812 that a continuous path was provided on the River Avon section. The Canal Company's committee of management reported in July of that year that:

It has been deemed advisable immediately to make a Horse Towing Path to the River Avon, under the Powers of an Act passed in the Year 1800. The Expenses of executing this Towing Path as well as of obtaining an Act of Parliament authorizing the same will be repaid to the Company, with interest, from the tolls of the Avon Navigation, and the Measure (which was unanimously approved of by the general meeting of the 1st Instant) is very desirable, even considered as a temporary Accommodation to the Trade, as it will materially facilitate the Passage of barges, and be a great Improvement on the present Mode of Towing by Men.

In *Waterways in the Making*, Edward Paget-Tomlinson points out that towpaths were owned by the Canal Company, were hedged off from adjoining land and were protected against trespass by various by-laws. Built about two feet above the water and surfaced with cinders or stone chippings known as 'raffle', the path was made to slope away from the waters edge so that horses had a better purchase when working the canal. The path occasionally changed sides to accommodate the requirements of local landowners, and when this happened the path was routed over what was known as a 'changeover bridge'.

Another common feature of canals were distance markers. Toll charges were levied on

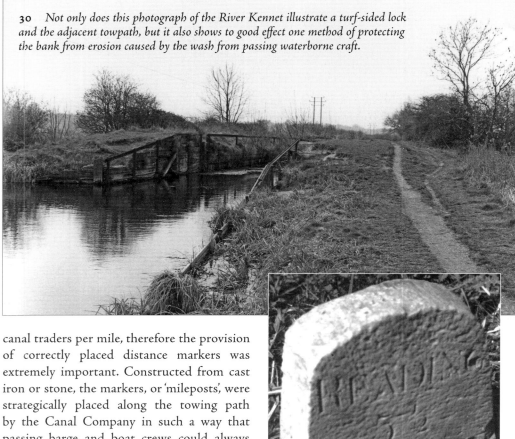

30　*Not only does this photograph of the River Kennet illustrate a turf-sided lock and the adjacent towpath, but it also shows to good effect one method of protecting the bank from erosion caused by the wash from passing waterborne craft.*

canal traders per mile, therefore the provision of correctly placed distance markers was extremely important. Constructed from cast iron or stone, the markers, or 'mileposts', were strategically placed along the towing path by the Canal Company in such a way that passing barge and boat crews could always see them, thus ensuring there would be no argument about paying the correct tolls on the pretence of not knowing the distance. Removing or defacing markers was considered a criminal offence. It is interesting to note that no distance markers before those used by the GWR have ever been found on the Kennet and Avon.

All these supporting structures and features were normally simple and straightforward to construct and locate. However, where long trunk canals such as the Kennet and Avon crossed valleys and rivers, or pierced high ground and hills, canal engineers had to construct technically complex and costly aqueducts and tunnels, with all their associated difficulties. In addition, a working canal required numerous structures

31　*A stone distance marker alongside the Navigation, near Reading.*

and amenities such as pound locks, bridges, wharves, and water supply arrangements, to ensure its effective operation.

In essence, canal aqueducts are artificial channels supported by bridge-like structures that convey the waterway across valleys, rivers,

32   *The Dundas aqueduct, named after Charles Dundas, the first Chairman of the Kennet and Avon Canal Company. It was built to take the canal across the Avon valley and thus enabled a lock-free nine-mile pound to be created from Bath to Bradford-on-Avon.*

flood arches, each with a span of 20 feet, also form part of the structure. Significantly longer but plainer in design, the aqueduct at Avoncliff has a single elliptical main span of 60 feet with two 34ft wide semi-circular flood arches flanking it. Both aqueducts have Corinthian-style entablatures below their parapet walls, and those at Dundas have exceptionally deep cornices that extend some four feet from the parapet. It has been suggested that this feature was not purely ornamental, but had

or other natural or man-made obstructions which could not be reasonably overcome by building locks. Seven aqueducts were constructed on the Kennet and Avon, five of which were minor structures enabling the canal to cross obstructions such as small rivers and streams. The remaining two were of a very different order and, as Paget-Tomlinson mentions, the magnificent stone aqueducts at Dundas and Avoncliff, which carry the canal across the River Avon and the valley in which it flows, are monuments to John Rennie's engineering expertise and to his mastery of classical design.

These substantial structures enabled the canal to cross the Avon valley twice, consequently allowing a nine-mile lock-free section to be created at the approaches to Bath. All Rennie's aqueducts were built of stone, and in *Navigable Waterways* (1969) L.T.C. Rolt notes that the Dundas aqueduct, the larger and more monumental of the two, measures 150 feet in length and has Roman Doric arches with twin pilasters either side of a 64ft wide semi-circular central arch. Two smaller parabolic

33   *A more recent photograph showing the canal crossing the Dundas aqueduct.*

34   *A stone post at the entrance to Avoncliff aqueduct, seen in the distance. It bears the scars and abrasions from more than 100 years of guiding grit-caked tow ropes as horses pulled boats and barges around the acute bend where the canal was routed over the aqueduct.*

the practical purpose of providing the masonry below with some protection from the weather.

Whilst aqueducts spanned valleys, areas of high ground across the canal's route were pierced by constructing deep cuttings or tunnels. Deep cutting was the cheapest method, and usually the canal engineer's first choice, although on occasion circumstances dictated that a tunnel had to be cut. On the Kennet and Avon, the 502yd long tunnel that cuts through a low hill near the ancient Savernake forest was a major feat of civil engineering, boasting one of the greatest cross-sections of any canal tunnel built during the 18th and 19th centuries. The portals and abutments of canal tunnels varied from plain brickwork to Doric pilasters and other fancy stonework, and Savernake, or the Bruce tunnel as it is more usually called, is of the more ornate John Rennie-style, with an imposing eastern entrance that is mainly stone-faced.

No towpath existed within the Bruce tunnel so barges had to be hauled through by hand, using a chain fixed to the southern wall. From a purely structural point of view, it is debatable whether this lengthy, fully brick-lined tunnel

35   *The eastern portal of the Bruce tunnel, above which can be seen the stone plaque inscribed in elegant Roman lettering. The plaque provides a testimony of gratitude to Thomas Bruce, Earl of Ailesbury, and his son, Charles Lord Bruce, for the consistent support they gave during the building of the canal.*

was really necessary, as a less costly deep cutting through the hill would have sufficed. But, the influential Thomas Bruce, Earl of Ailesbury, insisted that the tunnel be built as part of the compensation he was to receive for allowing the canal to cross his land. Although compensation to landowners often appeared to be excessive, this was the accepted way of doing business during the period that the Kennet and Avon was being constructed, and the Earl of Ailesbury was in fact one of the original members of the Canal Committee with a keen interest in the canal. In spite of the compensation he received, the Earl's involvement and contribution was obviously viewed with gratitude by other members of the committee, as a large stone plaque on the tunnel face proves.

Pound locks were so called because of the wider section of canal they incorporated. The 'pound' adjacent to each lock, from which water was drawn to fill empty lock chambers, was the means by which canal engineers overcame the

36   *A view of a lock gate and chamber showing the gate's general construction.*

37   *A sectional drawing of a typical lock and gate with all parts and structures labelled.*

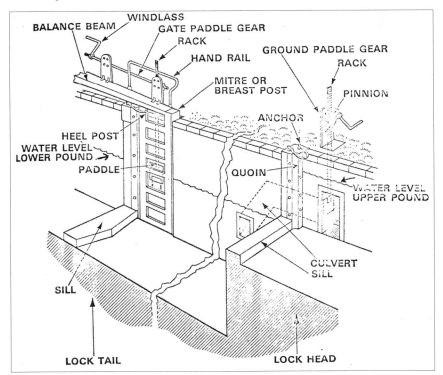

hills and valleys through which the waterway travelled. Pound locks are essentially chambers between two levels of water, closed by large oak gates at either end, which allow vessels travelling on a canal to enter at one level and, when the gates are closed and water let in to the lock, emerge at another level. Sometimes it was necessary to build flights of locks, such as those at Caen Hill, where 29 separate locks were constructed in such a way that 16 of them were close together and formed a flight stretching up the hillside from Foxhangers to Devizes. Space constraints meant it was not possible to build pounds in the normal way, so these essential elements of the lock system were constructed at 90 degrees to the canal line, forming large side ponds adjacent to each lock in the flight.

During periods when trading activity was high, and associated toll receipts justified it, the Canal Company would occasionally invest funds in improvements. These included the provision of gas lighting at the Caen Hill lock flight, using gas from the adjacent Devizes Gas Works. The Canal Committee resolved in their 1829 meeting that:

> No Barge or Boat be allowed to enter any one of the Devizes Locks after the Gas shall be lighted but on payment of one shilling for each Barge and six pence for each Boat which payment shall entitle the Owner to navigate his Barge or Boat through the said Locks so long as the Gas shall be lighted but no longer.

On the Kennet and Avon Canal there were 79 locks, with 31 rising to the summit level near Marlborough, and a further 48 falling along the canal's greater length towards sea level at Bath. When those on the adjoining river sections were also taken into account, the total number on the whole 86-odd miles of the Navigation amounted to 106 locks. Instead of being brick-lined in the normal way, the locks on the River

Kennet section were all turf-sided, with either brick or timber construction at their foundation levels, and originally designed to take the large sailing barges traditionally used on the river. These locks were subsequently reduced in length and width to bring them into line with those on the rest of the Navigation.

On wide canals, lock chambers were constructed to take barges that had a maximum length of 75 feet and a beam of just under 14 feet. Whilst these dimensions allowed one barge at a time to enter each chamber, the locks were wide enough to allow two narrow boats to be accommodated side by side, thereby speeding up passage arrangements. In accordance with normal practice, the locks on the Kennet and Avon Navigation were numbered as well as having names that were frequently based on local villages. Occasionally locks were named

38 *(Left) An aerial view of Caen Hill flight, showing two of its locks in use. As can be seen, two narrow boats can use a lock at the same time, although that was not the case for the wider Kennet barges.*

39 *(Below) A lock-side hut of the type that might have been used for temporary shelter by lock keepers and others when more permanent accommodation was not available.*

40 (Left) A lock keeper's cottage, near lock 44 at the top of Caen Hill lock flight.

41 (Below) This c.1900 photograph most likely shows the lock cottage on the Somerset Coal Canal where it joined the Kennet and Avon. The lock shown is of the narrow form used on the coal canal and is much smaller than the broad locks of the Kennet and Avon.

42 (Below) The underside of a brick-built hump-back bridge over the canal section. The towpath is clearly in evidence as it passes beneath the arch.

after particular people or after some other associated aspect or feature.

Some canal locks had huts alongside them, and these were used for temporary shelter by lock keepers who were responsible for more than one lock. Other locks had adjoining cottages that housed the lock keeper, together with his immediate family if he had one. Employed by the Canal Company, the lock keeper was there primarily to operate the locks and provide general assistance to crews so that boats and barges passed through relatively quickly. In addition, they were sometimes expected to collect tolls and to always ensure that no damage to lock structures occurred when the locks were being used.

If any damage did occur, or any of the by-laws of 1827 concerning the canal were broken, the lock keepers were expected to provide information on any offence committed to the local magistrates at the earliest opportunity so that swift punishment could be meted out. In an attempt to remove temptation and minimise unofficial deals between barge crews and lock keepers, the Canal Company supplied its lock keepers with coal and candles free of charge, which they were expected to only use for heating and lighting their cottages.

To ensure the continuing use of roads and paths affected by a canal's projected route, new bridges had to be constructed. In the case of the Kennet and Avon many of these bridges were hump-backed in form, and stone- or brick-built. On occasion, extremely ornamental stone as well as wrought-iron bridges were erected. Occurring in both rural and urban environments, they were used to carry footpaths across the canal and, like the Bruce tunnel, invariably resulted from compensation claims. Some of the most ornamental of these structures appear within the city of Bath, although one fine example was built in the open countryside near Wilcot in Wiltshire. Known as 'Lady's Bridge', this ornamental stone

43   An ornamental stone bridge that spans the canal at Wilcot and was erected as compensation for Mrs Susannah Wroughton, who was also paid 200 guineas and provided with an ornamental lake.

44   An ornamental iron bridge at Sydney Gardens in Bath, with a trip boat passing underneath. In addition to providing such bridges, the Canal Company had to pay a fee of 2,000 guineas for permission to route the waterway through Sydney Gardens.

45   An ornamental gate, again in Sydney Gardens, provided as compensation to an individual whose property was located alongside the canal.

footbridge was erected at the behest of a local landowner, Mrs Wroughton, who also insisted on a landscaped lake being excavated nearby.

Where the Kennet and Avon Canal crossed relatively flat countryside, numerous swing bridges, pivoting on massive iron ball race bearings, were erected. These were arranged in such a way that the weight of the perfectly balanced bridge was the only thing that kept the top and bottom cages of the ball race together. Rolt maintains that John Rennie pioneered this arrangement and that he was the first civil engineer to use ball races to assist in the operation of swing bridges across canals.

When the Kennet and Avon was a busy commercial waterway, and trade between Bristol, London and the many locations in between was thriving, there were around 120 wharves between Bristol and Reading. In order for cargoes to be worked effectively, many of these had wooden, or occasionally cast-iron, cranes erected on them. Where cranes were not available, wooden booms and lengths of

timber formed derricks that were used to load and unload cargoes. In the main, cities and bigger towns along the Navigation supported relatively large numbers of wharves, whilst fewer were constructed around villages, but it was not unusual for wharves to be provided in rural settings. Most of these are now quite isolated, but during the Navigation's prosperous years they would have been busy and important stopping points for canal carriers. Some isolated rural wharves were important enough to have their own horse railroads or tramway that linked them to nearby towns or villages.

Not every recorded wharf existed at the same time, although when the Navigation was being used commercially there were, for example, 18 wharves in Newbury and its surrounding area, with a further 17 in Reading. By the early 19th century wharves occupied the whole length of Reading's waterfront. At the other end of the canal, Bath could boast at least 19 wharves. Smaller conurbations along the canal generally had fewer, but the town of Devizes supported

46　*A swing bridge in operation, allowing a boat to travel through.*

47 *The only suspension bridge on the Navigation. Erected in 1845, the bridge was intended to link land at Stowell that was owned by Colonel Wroughton.*

seven at one stage. The wharves used by canal craft in the city of Bristol were mostly built for for sea-faring trading vessels, but elsewhere on the Navigation they were constructed primarily for river and canal work. On the River Avon part of the Navigation, coastal vessels sometimes traded as far upstream as Bath, a few of the smaller ones occasionally venturing onto the canal for short distances.

Sometimes owned by the Canal Company and sometimes by other companies or individuals, wharves were an essential element of the Navigation system. Larger wharves occasionally provided facilities such as warehousing and stabling for the towing horses, as well as supporting engineers, blacksmiths and boat and barge builders who all provided manufacturing and repair facilities. Artisans such as barrel makers, wheelwrights, and rope and canvas makers were also in evidence. Traders in bulk products like timber, coal, and

stone either located their yards at company wharves or used private ones; alternatively they constructed their own.

Although many of the wharves on the Kennet and Avon were relatively small, those in cities and towns such as Bath, Reading and Newbury were much larger, with an atmosphere and level of activity similar to that found in a medium-sized commercial seaport during the same period. The main wharf in Bath, known then and now as Broad Quay, was constructed in 1729, and some years after its completion, John Wood, in his *Essay Towards a Description of Bath* (1765), describes it as consisting of a large terrace extending to 483 feet in length by 97 feet in breadth, on which were sited four cranes and many buildings.

Owned by the Canal Company, Broad Quay supported numerous commercial establishments during its heyday. These included breweries, brass founders and ironworks, engineering

48   *A drawing of Crane wharf near Reading by N.H.J. Clarke, c.1920. It shows four Kennet barges alongside the wharf.*

49   *Hungerford wharf, c.1900, showing the warehouse building and wharf-side wooden crane. A boat appears to be just about to exit from the lock.*

50 *Bradford-on-Avon wharf, c.1900, showing the gauging dock with the quarter-ton stone blocks and crane used in the gauging process.*

and carriage works, boat and barge builders, rope yards, slaughterhouses, cabinet makers, carpenters and timber yards, as well as grocers, slate works and dye works. In the narrow streets that skirted the area were a multitude of inns and other establishments that sold alcohol, and the whole area must have been vibrant with activity and an intense mix of noise and odours, some of which might well have been noxious though unnoticed at the time.

Although large wharves tended to be situated within the bigger conurbations, and normally on the river sections, this was not always the case. The thriving wharf near the little village of Honeystreet was located on the canal in the depths of rural Wiltshire. The wharf and the undertakings that operated on it, together with most of the houses in the adjoining village, were owned by a partnership of three men. Samuel Robbins, Ebenezer Lane and Thomas Pinniger. These individuals developed a major business processing local hard and soft woods in their steam-driven sawmills, also importing significant quantities of more exotic timbers,

such as mahogany and teak, which were transported along the Navigation by barge from the partnership's storage areas at Bristol. Boat and barge building, the construction of various timber products, ironworking, the manufacture of fertiliser, tar oil, paint and cement, as well as slate and coal dealing, were some of the additional activities carried out at Honeystreet

51 *Taken in 1974, this photograph shows the inside of the building on Bradford-on-Avon wharf. The original roof construction can clearly be seen.*

wharf. Products were often distributed to the partnership's customers in its own fleet of barges and narrow boats.

Pound locks drew considerable quantities of water from the canal summit, and where wide canals were concerned this could mean the transfer of thousands of gallons of water for each lock opening. Canal engineers had to ensure that sufficient water was always available for the expected volume of traffic, by utilising either gravity feed or pumps. The river sections of the Navigation were self-feeding,

with various streams and springs providing the required water, but much of the canal was not able to benefit from such natural arrangements, and John Rennie and his engineers had to find other means of providing the necessary water.

The short summit pound had no water supply at summit level, although natural springs did exist at nearby Wilton, some 40 feet below the summit. Rennie initially diverted these springs so that they flowed into the canal at the lower level, then through a culvert to steam-driven pumps at Crofton, from where water

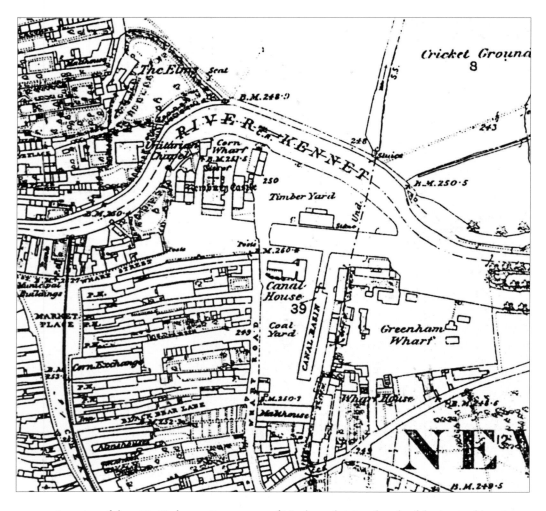

52   *A section of the 1880 Ordnance Survey map of Newbury showing the wharf, basins, and location of associated buildings and cargo storage areas.*

was pumped to the summit via a feeder channel. By 1836, however, it was clear that the springs could not provide sufficient water to cope with demand and a dam was built across a narrow valley at the summit level. The eight-acre area formed by the dam was then allowed to fill up with spring water to provide a reservoir capable of holding nearly seven million gallons of water. Natural rainfall and the replenishing water from the springs made up for evaporation and other losses from the reservoir.

The pumping installation at Crofton was extended and larger pumps were installed. Operated by two single-acting steam-driven beam engines, they were installed in a purpose-built engine house on which work first started in 1800. Many of the engineering parts used in the pumping installation, or station, were purchased from specialist suppliers and transported by horse and cart, whilst others were manufactured on site. In 1801 John Rennie purchased a Boulton and Watt beam engine that had originally been ordered for the West India Dock Company, and in 1812 he purchased another engine from the same maker. Once installed in the engine house, which provided the frame and structure to support the two beam engines and their massive 27ft cast-iron beams, the engines drew steam from three low-pressure haystack boilers

53 *Honeystreet wharf, c.1895, showing the various buildings, carts, and timber stacks ready for loading. The wide boat alongside is empty, but planks are laid along its length to enable crew members to move about more readily. The wooden crane appears to have been fitted with a wooden cover.*

and operated the pumping gear through parallel motion linkages and rods. Later the engines were converted to work on higher pressures, and the haystack boilers were replaced by three more efficient Cornish models. Much later, one of the Boulton and Watt engines was replaced and Lancashire boilers were installed in place of the Cornish type.

54 *The boat- and barge-building shed at Honeystreet, taken many years after all such work had ceased. During its peak this area would have been busy with numerous narrow boats and Kennet barges being built and launched sideways down well-greased wooden ways into the canal.*

55　*Crofton pumping station, with lock in the foreground. This photograph was taken sometime after 1958, when the boiler stack was reduced in height for safety reasons. This affected the natural draught to the boilers with a consequent adverse effect on boiler efficiency.*

Water supply problems still existed at the western end of the canal, however, and this prompted Rennie to purchase a River Avon corn mill at Claverton. After long delays resulting from issues surrounding water rights, he converted it into a pumping station. Unique to a British canal, the Claverton plant incorporated a 25ft wide breast shot wheel of more than 19ft diameter, and a 13ft diameter 'pit', or first motion wheel. It used a system of sluices and paddles that enabled water to turn the breast wheel which, in turn, rotated the pit wheel and moved a flywheel and crank via a follower. The resulting motion operated two cast-iron beams connected to lift and force pumps immersed in a water sump. It enabled water to be raised from the River Avon and back pumped up the lock flights on the canal, which at that point were located in close proximity to the river.

In addition to Crofton and Claverton, two further pumping stations containing beam engines were constructed in Bath. These installations were intended to provide water to the lock system in and around the city, but irregular traffic at the time, and objections and threatened legal action from local mill owners prompted the Canal Company to close the two pumping stations after only a few years' operation.

56 *(Above) The Lancashire boiler at Crofton pumping station with attendant raking clinker through the furnace fire-door.*

57 *(Right) The cylinder head room at Crofton pumping station showing steam piston cylinders and blocks, connecting and valve rods, and pivoted engine beams above.*

58   (Above) The Claverton pumping station water wheel and associated gearing. The wheel is breast shot from the millpond.

59   (Right) The pit, or first motion wheel, at Claverton pumping station. Turned by the wooden breast shot wheel, this cast-iron wheel operated a train of wheels that in turn operated the pumps by means of a crankshaft and two pivoted beams.

60   All that remains of the smaller pumping stations built near Bath is this ornamental chimney located near Abbey View lock.

5 | WIDE AND NARROW, SLOW AND FAST:
THE BARGES AND BOATS

During the extended period when the rapidly growing network of inland waterways in England allowed waterborne trade to spread between many centres of population, a number of different barge forms were developed in response to local conditions and economic factors. As Tony Condor has pointed out, in *Canal Narrow Boats and Barges* (2004), it is likely that a handful of distinct types of barge significantly affected the evolution of all other types. The sail-powered primary types would have been linked to large rivers and estuaries and are likely to have included the sailing flats of the Mersey, the keels of the Tyne and Humber, the keels and wherries of the Norfolk broads and rivers, the Thames and Medway sailing barges, and the trows of the River Severn and River Wye.

Sails were not so useful in the restricted waters of small rivers or canals and as the canal network progressed it soon became evident that different forms of motive power were required. As a consequence, 'dumb' barges that could be towed or manoeuvred with sweeps or poles were developed from the primary sailing types, although a short, simply rigged mast with one or more square sails was sometimes retained to assist with powering the barge when conditions allowed. A further barge type evolved that had a similar length but a much reduced beam, to enable working on the narrowest canals. Such craft were known as 'narrow boats' or, less frequently, 'long boats', and were used on just about every canal in England.

The trading barges used by carriers on the Kennet and Avon Navigation were of these derived types, consisting of large barges known as 'Kennet barges', narrow boats of the usual shape and form, and craft known as 'wide boats' or 'mules', which looked like narrow boats but were somewhat wider in the beam. In addition, a number of passenger and other more specialist craft evolved that were used in response to particular needs or economic factors.

Although the different types of barge were often distinctive, the shape and form of the Kennet barge, with its shapely transom stern, rounded turn of the bilges, and bold shear, was particularly so, especially for a dumb barge destined to spend at least part of its working life in a canal environment. It is therefore interesting to speculate on where such a design might have come from, and what influences could have prompted its widescale use on the Kennet and Avon. It should be remembered that the highly conservative Victorian and Edwardian men, whose living revolved around river and canal trading, were unlikely to have favoured a barge form that differed greatly from those with which they were familiar, unless of course there were good practical reasons to do so.

61　*This photograph, taken c.1892 near Limpley Stoke bridge, shows the Kennet barge,* Pearl. *Built at Honeystreet for the United Alkali Company, the* Pearl *is probably on her delivery voyage to her new owners at Bristol.*

As we have already seen, extensive trading took place on the rivers that formed the two ends of the Kennet and Avon Navigation in the period before the canal section was opened. In the east, on the River Kennet, the massive Newbury barge was a common sight. Owing its origins to the so-called 'Western barge' that had been used on the upper stretches of the River Thames for hundreds of years, the Newbury barge was more than one hundred feet long, with a beam of around seventeen feet. Box-like, with a flat, keel-less bottom and upward-sloping square-cut ends known as 'swims', these craft were capable of carrying up to 110 tons of cargo. Powered by a single square sail set on a short, centrally-placed mast, Newbury barges had large holds that took up the majority of the internal space; in an area at the stern wooden hoops supported a canvas shelter for the crew, covered wagon-style. Much too long and wide to operate on the canal and manoeuvre through

its locks, these barges eventually became totally redundant when the large turf-sided locks on the River Kennet were reduced in size. With no obvious links or similarities, it is highly unlikely that Newbury barges had any influence whatsoever on the Kennet barge form.

By contrast, barges used on the River Avon were generally smaller, and in practice were likely to have been Severn trows – pronounced like 'crows'. These unique sailing barges were traditionally used on the River Severn and River Wye and would originally have carried one or more square sails set on a single mast. Later, those used on the River Severn and in the Bristol Channel were fore and aft rigged as sloops, or even ketches, with the addition of a short mizzenmast. Having an open hold, with decks at bow and stern that housed crew accommodation as well as storage space, trows were most conveniently used in the non-tidal parts of the River Avon between Bristol and

Bath by 'cutting down'. This process involved removing all masts, spars, sails, rigging and other associated equipment, thereby converting the vessel into a dumb river barge. At least one trow in a cut down condition was still employed on the River Avon as late as the 1930s.

Bristol Channel traders who had travelled down the River Severn in their trows with cargoes destined for Bath would frequently leave out the middle man and, instead of trans-shipping cargoes at Bristol to river barges, they would work their sailing vessels up river to unload them directly at Bath. In this way they could make significant savings on handling and haulage costs, particularly when bulk cargoes such as South Wales coal or Droitwich salt were being transported.

During this period, therefore, both cut down and sailing trows would have been a common sight on the river, as well as alongside the quays and wharves at Bath.

Contemporary paintings of Bath and the River Avon sometimes depict another type of small barge with rounded ends. These do not appear to be trows, and whilst it is possible that they illustrate another local type, it is more likely that artistic license has been displayed in depicting them. In any case, the general shape of the Severn trow was such that the Kennet barge form is more readily attributable to it than any other vessel used during the period.

Both Kennet barges and trows had rounded bows with straight-sided hulls and flat bottoms,

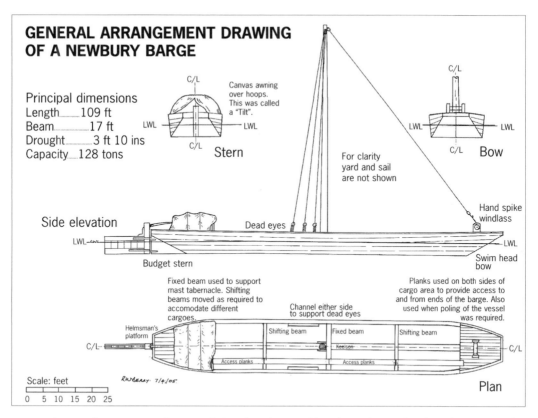

62 *A general arrangement drawing of a Newbury barge. Although not shown, these huge barges carried a single square sail on their short masts and had minimal running rigging.*

all set on frames that were tightly curved at the turn of the bilge. Short fore and aft decks, with a large hold between that was traversed by a number of substantial wooden beams, was also a feature of both types. Whilst most of these hold beams could be moved to accommodate different cargoes, the foremost one was always fixed in position and supported a mast tabernacle at its mid point that extended down to the massive keelson, the main fore and aft strengthening timber.

Unlike that fitted in trows, the Kennet barge tabernacle was not normally used to support masts and sails but to house a simple wooden derrick for working cargo when no external crane was available. The tabernacle could also have been used to house a short towing mast, but barges on the Kennet and Avon were rarely towed in this fashion, the deck windlass, bitts and hawsehole for guiding the towrope being used instead. In addition to these similarities, two visual aspects confirm a close relationship between Severn trows and Kennet barges: the bold sheer line and distinctive transom stern shaped like a 'D' on its side were prominently exhibited by the wooden hulls of both types.

Although unique to the South West of England, the Severn trow had always been a popular and successful form of working craft, and over the long period during which it was in use a number of different variations developed in response to the requirements of trade and other local factors. These variations included smaller, shallower forms for use in the upper waters of the River Severn and on the River Wye, and larger, deeper-hulled types for use in the lower parts of the Severn and in the Bristol Channel. Many examples of this latter type were even capable of coastal trading and operating in what were frequently harsh sea conditions.

One of the more important trades down the Severn to Bristol was salt from Droitwich in Worcestershire. As Colin Green has pointed out in *Severn Traders* (1999), the

63   (*Above*) *A derelict Severn trow at Dundas wharf in 1956. One of the few trows then remaining, this one has ventured a long way up the canal. Eventually rotting away at her moorings, the vessel shows the marked similarity between the trow and Kennet barge types.*

64   (*Right*) *A photograph, c.1920, showing a Kennet barge at Honeystreet wharf. Behind the barge are a number of narrow boats and a second Kennet barge being either loaded or unloaded near the building sheds. The rounded transom top on the first barge indicates the 'wich barge' influence.*

salt trade had increased to such an extent by 1771 that a seven-mile long barge canal was opened to connect Droitwich with the River Severn more directly. The barges used for transporting salt, known as 'wich' barges, were in fact small trows with more shallow, elliptical 'D'-shaped sterns than was normally the case with the larger trows. In contrast to the drab exterior normally associated with trows, these barges were also provided with brightly coloured carvings at the stern and quarters. But the elegant wich barge bore an even closer resemblance to its later sister, the Kennet barge, than did other trow forms, and appears to have exerted a strong influence on the development of the latter, particularly where hull shape was concerned.

Both Kennet and wich barges were less cumbersome in appearance and more shapely than normal trows. These differences reflected the need for larger trows to be capable of making coastal passages and dealing with stormy conditions, whilst both Kennet and

65 *Taken at Bristol in 1934, this photograph shows Kennet barges* Celtic *and* Ajax *alongside the Severn trow,* Aurora. *The trow has been cut down, with mast and rigging removed, and all three barges are being used as lighters in Bristol docks. The similarity between Kennet barges and trow forms is evident.*

the much older wich barges would normally only have been used on river and canal. The wich influence on Kennet barge hull form would most likely have spread up the River Avon from Bristol, as well as being focused to a lesser extent on Bath. At both locations it is probable that barge crews and commercial carriers, together with boat and barge builders, studied and were impressed with these attractive yet work-like vessels.

Once the whole of the Kennet and Avon Navigation was opened for trading, existing trows in a cut down condition would, in all likelihood, have also been used on parts of the canal section. In practice, however, the restrictions in width that canal operations created would have meant such craft were not entirely suitable, and their sphere of use would have been limited. It was thus only natural that barge builders would soon start building vessels that better suited the requirements of

traders who wanted to use both river and canal sections of the Navigation seamlessly. Purpose-built dumb barges, probably based on the wich barge form (around 70 feet long, up to 14 feet wide, and capable of carrying about 60 tons), would have increasingly been seen. Initially known as 'Kennet and Avon' barges, a name that was soon shortened to 'Kennet' barge,

67 *(Right) The John Gould-owned narrow boats,* Colin *and* Iris, *approaching Bulls lock in 1950. With the advent of reliable engines, this arrangement of two boats working together become common practice. Within this pair,* Colin, *an ex-horse boat, is the butty, and* Iris *is the motorboat providing the power.*

66 *The recently launched Kennet barge, Diamond, shown at Honeystreet wharf around 1930. A boiler, possibly for delivery along the Navigation, lies on the wharf, and behind the wooden crane is a long-wheeled carriage for transporting logs of timber.*

these became the most popular type of larger vessel to be used on the Kennet and Avon.

Owing their development to the earliest forms of canal boat, such as those used on the Bridgwater Canal, all commercial narrow boats, no matter where they were used, were of a similar design. Able to carry up to around twenty tons, narrow boats were suitable for moving a range of cargoes but were not entirely suitable for some types of bulk cargo such as timber, and in any case their capacity was much less than that of a Kennet barge. The narrow boat's great strength, however, was its ability to travel on all parts of the inland waterway system, and with its 70ft length and 7ft beam it could use wide canals as well as the more narrow ones where barges were unable to venture.

Shaped like a long box, with rounded bow and stern and slightly flared sides, the greater part of a narrow boat was given over to cargo space, although there was a short, decked area in the bow and an after cabin to accommodate the crew. In the larger barge, protection for the cargo was provided by simply covering it with a tarpaulin, The arrangement in a narrow boat, however, was more complex and involved a shaped bulkhead, called a 'cratch', that was situated near the bows, and various wooden members known as 'stretchers' and stands placed between the cratch and the cabin front. Over the top of these structures was placed a top plank, and tarpaulins called 'side cloths' were subsequently draped over this to form what resembled an elongated tent.

The crews of narrow boats on many canals lived on board and were frequently made up from members of the family of the boat owner, or 'Number One' as the skipper was called. Whilst such boats undoubtedly traded on the Kennet and Avon, it was more usual for narrow boats there to be crewed by individuals, or workers who did not normally live on board. This arrangement also applied to Kennet barges and, although basic accommodation was provided on board, many crews worked a system of stages, changing boats in opposite directions and spending the nights in local inns or hostels where stabling for horses was provided.

As they were purely working boats and not family homes, long-distance narrow boats owned by the carrying companies on the Kennet and Avon displayed very little, if any, of the elaborate decoration that became fashionable on some canals after about 1875. Although barges were exempt, after 1871 narrow boat owners who did live aboard their boats had to be registered by a Municipal Authority such as Bath Corporation, requirements for this being laid out in the Canal Boats Acts of 1871 and 1884 respectively. Registration authorities had powers to inspect and impose penalties on boat owners if the regulations included in these Acts were being disobeyed.

In the main, both barges and narrow boats were moved by towing horses. As Paget-Tomlinson relates, to move a loaded barge or narrow boat from a stationary position a horse had to move forward against its collar until it had broken the craft's inertia. The first few steps were difficult, but once the barge or boat had gathered momentum the horse could step forward more rapidly, and towing gradually became easier as the great bulk of vessel and cargo glided through the water. Many carrying companies, small traders and individual Number Ones owned their own horses, but these animals could be conveniently hired when necessary.

Horses worked all day and so had to be fed and watered en route. This was done using a nose tin secured with a strap that passed around the horse's neck. Often muzzled to discourage grazing on the way, horses were also blinkered so they would not become distracted. Horse harnesses and gear were relatively simple, and the tow line was so arranged that it could be readily cast off in the event of a barge yawing

68 *Two motorboats at Reading. This photograph clearly shows the cratches, stretchers, and stands that were an integral part of a trading narrow boat. These boats are quite well decorated so may be visitors to the Kennet and Avon.*

69 *A view of a river section of the Navigation showing bow-hauling in practice.*

away from the towpath and threatening to pull the horse into the canal. Like most other canal proprietors, the Kennet and Avon Canal Company did not allow cargo carriers to travel on the Navigation during the hours of darkness, and enforced this ruling by having lock gates secured by chains in the closed position. During the night, therefore, horses had to be stabled.

Although a towpath was built along the whole length of the canal section, the arrangements along the river sections were not initially as good, and towpaths were not provided there for some years after through-trading had commenced. Due to this, barges and narrow boats travelling on the river sections had to be bow-hauled using relatively large teams of men, especially when travelling upstream against the current, when controlled drifting was not possible.

Describing these teams in *The Making of the Middle Thames* (1977), David Gordon Wilson says that the hauliers, or 'halers' as they were more commonly known, were often destitute

70   (Above) Another view of the narrow boats Colin *and* Iris *arriving at Newbury in the winter of 1949.*

and ragged men who waited at town wharves and fought for the opportunity to earn a few pennies by towing craft up to the next reach. Working up to their waists in the river, these tough individuals had to haul waterlogged cables that were often stiff with ice or contained river grit that mercilessly abraded their hands, arms and shoulders. This practice appears to have continued occasionally on the Kennet and Avon even after the towpath was eventually extended along the river sections, the lower cost of bow-hauling being enough incentive for halers to be used instead of horses.

With advances in technology, mechanical propulsion became possible on the canal network. Although the barge *Comet* was fitted with a steam engine when built in 1890, steam engines were rarely used in practice on the Kennet and Avon because of their general unreliability. The later oil engines, whilst often installed in narrow boats, appear never to have been fitted to Kennet barges. This may have been due to the existence of the massive internal keelson and the huge externally hung rudder, both of which would have made the fitting of a stern tube and propeller especially difficult unless extensive and costly structural changes were carried out. Once small oil engines had

72   *A wide boat shown alongside the wharf at Pewsey in 1975. The unusual looking craft emerging from underneath the bridge is the paddle boat,* Charlotte Dundas, *that was built by the Kennet and Avon Canal Trust and operated as a trip boat.*

become more efficient, in the early 20th century, many boat owners changed from horse-drawn narrow boats to what were by then known as 'motorboats'. Initially, wooden horse-drawn boats were converted to house engines, but this arrangement was not particularly successful and steel motorboats were introduced. The old wooden boats were retained as dumb boats, or 'butties', that were towed behind the powered boat, thus forming a working pair.

As an alternative, wide boats that were locally known as 'mules' were sometimes used. These were built on narrow boat lines, but had a beam of around ten feet at the gunwale, tapering to about seven feet across the bottom, an arrangement that appears to have been initiated in an attempt to decrease water resistance. With a length of about seventy feet, most other features of the wide boat were the same as those of a narrow boat, but a wider beam allowed them to carry up to fifty tons.

Although a number of factors, such as the quantity and type of goods being carried, the number of horses employed, and the number

71   *(Left) A narrow boat pair showing the way tarpaulins were used to cover cargoes.*

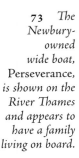

73 The Newbury-owned wide boat, Perseverance, is shown on the River Thames and appears to have a family living on board.

of locks to be passed, would appear to have affected journey times of both boats and barges, in practice times on the Kennet and Avon appeared to be relatively consistent. This is evident from an investigation carried out by the Canal Company, whose findings were presented to the proprietors in 1818. The report stated that:

> On examining the books kept by the tonnage clerks, it appears that the number of small boats, now employed on the canal, exceeds two hundred, and that there are about seventy barges of sixty tons each – that the annual average time of the barges of the two most considerable traders, passing along the Kennet and Avon Canal, between Bath and Newbury (being a distance of fifty-seven miles) has not, in the last four years, varied more than about one hour – and that such average has been three days and nine hours. It also appears that since the last improvements have

been effected, on the River Kennet and the River Avon, no material impediments have occurred on those navigations. The evidence of these facts more than ever justifies your Committee in stating, that the uncertainty and delays, which have so frequently occurred in the conveyance of goods between the cities of London and Bristol, have arisen on the Thames, and been occasioned by the natural and insurmountable obstructions in the navigation of that river; and the same facts will also shew to the Proprietors which might be done under the advantage of a canal made to avoid these obstructions, independently of what would arise from fly boats being enabled to pass from Bristol to London, which they cannot do so long as the River Thames forms part of the communication.

Apart from the eventual introduction of oil engines and the use of iron and then steel

74   A fully loaded horse-drawn narrow boat in the Bath Valley. The photograph was probably taken in the late 19th century.

75   The motorised tar barge, Derby, photographed in 1955 in Swineford lock on the River Avon. The barge was used daily to collect 50 tons of tar from Bath gasworks for delivery to the Bristol and West of England Tar Distillers Ltd at Crewes Hole near Bristol.

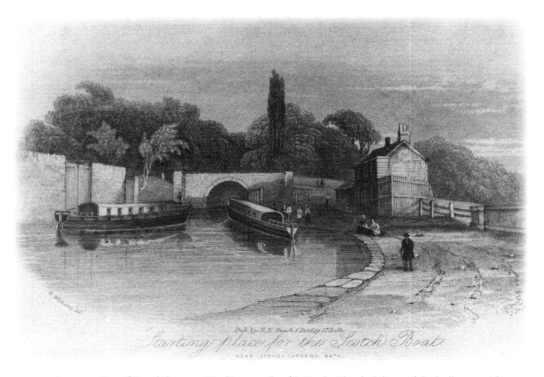

76 *An engraving of Scotch boats at Darlington wharf in Bath. The sleek lines of the hulls are evident, as too are the cabin arrangements that such boats boasted.*

78 A Great Western Railway working flat. Pictured near Devizes in around 1910, the form of these maintenance boats changed little over the Navigation's long history.

for hull construction, the barges and narrow boats involved in general trading on the Kennet and Avon were subject to very little change throughout the course of the 137-odd years that the Navigation was operational. However, a number of additional craft were developed in response to specific needs. These included a small number of motorised steel tar barges that

77 (Left) During the First World War the old workhouse at Avoncliffe became a convalescent hospital for injured soldiers. The horse-drawn barge, Bittern, was regularly used to transport patients to and from a public house in Bradford-on-Avon, presumably as part of their convalescence.

were used regularly to collect tar from Bath Gasworks and transport it to the tar distillers at Bristol, and boats that specifically provided a regular passenger service along the canal, in contrast to the occasional narrow boat that provided a similar service.

Generally known as 'packet boats', or 'fly-boats' where the fastest were concerned, these distinctive-looking craft provided a regular service between Bath and London, via the River Thames. The packet boats that operated between Bath and Bradford-on-Avon, however, were called 'Scotch boats', because the prototype had been brought from Scotland at the joint expense of several canal companies all interested in the design. With long, narrow hulls similar to a narrow boat, fly-boats had a more hollow stern and finer lines. Sufficient cabin space was provided for both first- and second-class passengers, and boats on longer runs had sleeping and eating facilities on board. According to Clew, the Scotch boat form of fly-boat sometimes even provided light entertainment and a string band for the enjoyment of its passengers. Usually pulled by two horses, each of which had a rider, the fly-boats were stopped at regular intervals en route so that the horses could be changed. Given priority on all parts of the Navigation, the swiftest of these passenger boats were drawn by teams of galloping or cantering horses and were often allowed to pass through locks during the night so as to maintain their scheduled passage times.

The Canal Company also used certain specialist boats for maintenance purposes. In the main these boats had simple raft-like hulls with dredging equipment or lifting facilities on board, or were fitted with equipment for breaking ice. One maintenance boat type was particularly noteworthy, however. During a major dredging programme early in the 20th century, the Canal Company purchased a number of wooden boats that had been used on both the Monmouth and Brecon and the Swansea Canals. Known, not surprisingly, as 'Welsh boats' and capable of carrying around twenty tons, these craft were shorter, but broader in the beam, than traditional narrow boats. Few carried cabins and all were double-ended so that the rudder could be transferred from one end to the other, making the boats more manoeuvrable than would otherwise have been possible. These craft were subsequently used on the Kennet and Avon for loading with mud during programmed dredging operations, and were employed in that important task for many years afterwards.

6 | THAT RIBBON OF INDUSTRY:
TRADE AND TRADERS, CARGOES AND CARRIERS

Now that the Navigation was completed it had to be paid for, and those who had provided funds naturally expected a return on their investment. Consequently, the cost to carriers of having a through route from Bristol to the River Thames and on to London was an increase in the tolls that had applied when the Navigation had only been partly open. The Acts of Parliament that enabled canals to be built had stipulated that companies who built and operated them could only generate income by charging tolls. In an attempt to make canals more competitive, however, amendments were introduced that allowed the companies to operate as carriers themselves. These changes enabled the Kennet and Avon Canal Company to become involved in carrying, and much later in providing passenger services, but in the main its role continued to be one of toll administration and collection, navigation management and maintenance, and the provision of warehousing and support facilities. Goods were transported in the barges and boats of large and small independent traders.

Different goods attracted different tolls, with costs per ton being lower for bulk items such as coal and stone, whilst finished goods and more highly priced general merchandise cost more to transport. The maximum toll which could be charged for each class of goods was specified in the appropriate Act and could not be increased

without further Parliamentary approval, as Hadfield points out. But growing competition from gradually improving land carriage and from the coastal trade, which was becoming safer due to more effective convoy arrangements and the effects of the British Navy's close blockade of the French coast, tended to keep inflationary increases low. Gradually expanding railway competition, as well as pressure from industrialists and large carriers to reduce costs, also had an effect. All these factors combined tended to force the Canal Company to keep its tolls at a competitive level.

Tolls were charged on the basis of tons per mile. Each carrier was required to produce a written note describing the cargo and indicating its weight. This note had to be endorsed by a toll-collector, who had the authority to question the accuracy of the carrier's estimate, especially if he suspected that an attempt was being made to pay a lower toll rate. When this occurred the toll collector did his own check of the tonnage being carried using a process known as 'gauging'.

Before barges or boats could be used on a particular navigation they were taken to a gauging dock in an empty condition. Here the dry inches, or the distance from the top of the gunwale to the water surface, were measured. Quarter-ton weights were then gradually added by placing them in the vessel's cargo space, and the measurement between gunwale and water taken on each occasion. These measurements, which recorded dry inches against cumulative onboard weights, were then recorded in tabular form in ledgers that were sometimes retained at lock cottages, or more usually in the tollhouses that existed at all the larger wharves. The depth at which a barge or boat was floating at any given time was determined by placing a gauging device in the water alongside it. This hollow metal tube, with an open bottom and a protruding metal arm near the top, rested on the vessel's gunwale and acted as a constant index point when the gauging rod was inserted in the water to a specific depth. Inside the tube was a float and a rod marked in feet and inches which rose up the tube and indicated the freeboard at that particular point. Readings at four to six different positions were taken and averaged before being compared with the figures in the company ledger for that particular barge or boat. Once this was done the distance

the barge or boat had, or was about, to travel was taken into account and the relevant toll rate calculated. The system worked reasonably well but it was far from foolproof, particularly on long trunk navigations such as the Kennet and Avon. The waterway was used by many different types of craft carrying mixed cargoes which, in the first instance, were only subject to the carrier's declaration as to weight and content. Gauging was a source of conflict between Canal Company employees and traders for much of the Navigation's history.

As Clew points out, one of the most important effects of the canal's completion was the increased prosperity it brought to the towns and villages along the route. Various goods and provisions could now be brought in quantity from previously barely accessible locations, and a much wider market for local produce became more easily available. A barge could generally carry some forty tons of cargo, compared to the two tons that could be transported in a horse-

TONNAGE RATES.

	s.	d.	
For all Hay, Straw, Dung, Peat, and Peat-ashes, and all other Ashes used for Manure, Chalk, Marl, Clay, and Sand, and all other Articles used for Manure and for the Repair of Roads	0	1	per Ton, per Mile.
For all Coals, Culm, Coke, Cinders, Charcoal, Iron-stone, Pig-iron, Iron-ore, Copper-ore, Lead-ore, Lime, (except used for Manure,) Lime-stone, and other Stone, Bricks and Tiles	0	1½	ditto. ditto.
For all Corn and other Grain, Flour, Malt, Meal, Timber, Bar-iron, and Lead, (except such Corn, and other Grain, Flour, Malt, and Meal, as shall be carried Westwards, on such part of the Canal as shall be situate between the Town of Devizes and the City of Bath)	0	2	ditto. ditto.
For all Corn, and other Grain, Flour, Malt, and Meal, which shall be carried from the Town of Devizes to the City of Bath	3	0	per Ton.
For all Corn and other Grain, Flour, Malt, and Meal, which shall be carried Westwards on any part of the said Canal between the Town of Devizes and the City of Bath, and shall not pass the whole Way between Devizes and Bath	0	1½	ditto, per Mile.
For all other Goods, Wares, Merchandize, and Commodities whatsoever, in respect of which no Toll, Rate, or Duty is hereinbefore made payable	0	2½	ditto. ditto.

And so on in Proportion for any Quantity greater or less than a Ton, and for any Distance more or less than a Mile.

79 *Tonnage rates, c.1810, which were applied to the main cargoes carried on the Navigation.*

DEVIZES WHARF.

THE following RATES and SUMS of MONEY are to be paid at this Wharf, from and after the 31st of December, 1810.

FOR WHARFAGE.

	£	s.	d.
OF Coals, Culm, Limestone, Clay, Iron, Iron Stone, Lead Ore or any other Ores, Timber, Stone, Bricks, Tiles, Slates, Gravel, Hay, Straw, Corn in the Straw, or Manure whatsoever, which shall be placed on this Wharf, and continue thereon for any time not exceeding One Calendar Month, —————— per Ton,		0	3
OF any other Goods, Wares, Merchandizes or Commodities whatsoever which shall be placed on this Wharf or continue thereon for any time not exceeding six days, —————— per Ton		0	2
AND for every Day above either of the above Periods —— —— per Ton	0	0	1

N.B. The above Rates are allowed by the Act of Parliament.

FOR CRANAGE.

	£	s.	d.
OF Timber, Stone, and all other heavy Articles —— —— per Ton	0	0	4

FOR WEIGHING GOODS
NOT WAREHOUSED.

	£	s.	d.
ALL heavy Goods, weighing five hundred weight, or upwards, —— per Ton	0	0	6
ALL small Articles, under five hundred weight, —— at per Article,	0	0	1½

FOR WAREHOUSING,
(WEIGHING INCLUDED).

	£	s.	d.
ALL Goods, of every description, which shall be deposited in the Public Warehouse, and continue there for any time not exceeding One Calendar Month —————— at per Hundred Weight	0	0	0½

THE Proprietors are not answerable for any Goods deposited in the Warehouse or on any Part of the Wharf, which may be lost or damaged by Fire.

THE Proprietors direct that every Bargemaster or other Person, having the charge of any Boat or Barge navigating the Canal, shall deliver a particular account of the Quantity and Weight of their respective Loads to the Wharfinger, either before or immediately after loading or unloading the same ; and that all Boats and Barges entering this Wharf, are to be placed and arranged from time to time, under the Direction of the Wharfinger.

N. B. No Fire will at any time be permitted to be lighted in any Barge lying under or contiguous to the Warehouse.

By Order of the Proprietors.

28th December, 1810.

R. Sweeper, Printer, Devizes.

80   *An example of the type and value of charges the Canal Company applied for the use of wharf facilities in 1810.*

81   *Bradford-on-Avon wharf in about 1920. The two narrow boats on the right appear to be moored in front of the lock top gates, making entry or exit by other boats difficult. The gauging dock and its crane and weights is on the left.*

drawn wagon travelling on a surfaced road. Where the transport of goods between Bath and London, for example, was concerned, this equated to a carriage cost by barge of under £3 per ton, and more than £6 for road transport, but a dramatic decrease in land transport costs after the Napoleonic Wars reduced this margin.

During its first forty years of operation the Kennet and Avon continued to be a successful trading concern, although as Hadfield notes in *British Canals* its dividends were small as the cost of upkeep was high. By 1823 some 127,000 tons of cargo were being transported, and the Navigation was sufficiently prosperous for toll receipts to increase to a value of nearly £38,000, enabling the company to pay an increased dividend to its shareholders. Some years later nearly 342,000 tons of cargo were being carried, an amount that was more than five times greater

than that transported during the same period on the more northerly, although similarly routed, Thames and Severn narrow canal.

The prosperous years for the waterway were between 1810 and 1852, and during that period a number of hitherto small traders were able to expand and build relatively large businesses, retaining fleets of barges and narrow boats and employing their own staff to operate them. As they frequently owned or leased wharf facilities, the more entrepreneurial of these men, either singly or in partnership, often diversified. For example, they might establish timber conversion and supply arrangements using steam-powered machinery or construct boat- and barge-building yards or ironworking and wheelwrights' shops. At least one family concern, the Falls of Burbage, who were tenant farmers before becoming wharfingers at Burbage wharf, also

82 (Left) The narrow boats Caroline and Apollo wait for the lock to empty before ascending Caen Hill. Both boats are loaded and the boatman has his family on board. The horse waits patiently for the lock to fill, c.1870.

83 (Right) A narrow boat waiting at a lock whilst the two women converse, c.1920.

84 (Below) An empty narrow boat entering the bottom lock of Caen Hill flight after descending from Devizes, c.1850.

undertook steam threshing and ploughing work for other local farmers, utilising their fleet of powerful steam traction engines.

Others produced various types of fertilizer to augment the increasingly scarce natural manure which was required by the local agricultural community. This so-called 'artificial manure', consisting of organic matter such as crushed bone and wood waste to which both alkalis and acids were added, was produced in pits known as 'manure plants'. All the items required, including cargoes of bone, limestone and carboys of acid, were delivered along the Navigation, which was also used to transport the finished product. Where manure plant and timber conversion existed in close proximity, the waste from timber conversion activities was frequently recycled to provide part of the base ingredient in the artificial manure production

process. Waste from timber conversion was also used by other manufacturers, waste oak chips being transported along the canal to the village of Semington, for example, from where they were taken up the adjoining Wilts and Berks Canal to the meat processing works of Harris of Calne and used in the smoking of bacon. Later, the carriage of passengers, together with the necessary support arrangements, was an additional area for diversification where the larger of these businesses were concerned.

Some of the wharf operators were relatively enlightened employers for the period. Robbins, Lane and Pinniger of Honeystreet wharf built subsidised housing for their 136 employees, provided a meeting hall and certain leisure facilities, and arranged for payments to be made to workers who had become incapacitated due to work-related accidents or ill health. Every

85 *The Kennet barge,* Unity, *moored at Lower Wharf Devizes in 1810. Loaded with empty acid carboys, the* Unity *is returning its cargo from the artificial manure plant at Honeystreet to the United Alkali factory at Avonmouth.*

86 *Kintbury wharf in around 1898. Logs are being converted into deals and fencing stakes, using machinery driven by a steam traction engine. The finished products will subsequently be shipped along the Navigation.*

employee of the partnership had to pay into a sickness fund to enjoy fully its benefits, each man paying a membership of 1s. a month, and every boy half that amount. In addition, doctors' fees of 3s. to 4s. a year had to be paid. The rules and sickness benefits available were laid out in a document entitled 'The Honeystreet Wharf Accident and Sick Fund', but the partners were only prepared to take their generosity so far and the 1896 edition of the document included a number of what now would appear to be fairly draconian rules:

> Anyone being unable to follow his daily occupation through accident or sickness arising from drunkenness, fighting, running, jumping, gambling, or while suffering from venereal disease, shall not be entitled to any relief from this fund, nor on account of insanity or blindness

> … If any man or boy, while receiving sick pay, be found doing any kind of work whatsoever, or be seen out of doors after half past five in the evening during the seven winter months, or half past eight in the evening during the five summer months, he shall be disentitled to any further pay during his illness.

The Honeystreet Wharf Accident and Sick Fund provides an interesting insight into the way 19th-century employers viewed their workers. Clearly the fund arrangements were designed to retain staff and reward commitment in an age when social benefits were few, and being unable to earn a living was a major, if not disastrous, difficulty for most workers. However, it is evident from the wording, intention and constraints included in the Honeystreet document that even the more

87   *Hungerford wharf in 1890. In front of the wooden crane is a large log and sawdust pile. It is not clear what work is going on, but a narrow boat and crew wait at the wharf and appear to be taking a rest from loading or unloading their cargo.*

88   *The Kennet barge, Unity, at Honeystreet wharf. The shapely transom stern and bold shear of the barge's bulwarks are shown to good effect in this photograph.*

enlightened employers used a management style that could at best be described as one of benevolent autocracy.

By and large, however, employment in the businesses along the waterway was not too bad, with skilled workers such as boat builders and carpenters earning around 30s. a week in the latter part of the 19th century. This wage could be increased if additional work was carried out: at Honeystreet, for example, an extra 2s. per week could be earned if evening and night-time work manufacturing wooden gates was undertaken. Wage rates of this level are quite good compared with those of similar occupations elsewhere in late 19th-century Britain. There are indications, however, that the weekly remuneration for employed barge masters was less than that received by skilled tradesmen. But if an employer paid for horse stabling and other expenses, and housing was provided, then his was still a reasonable income. It should be noted

that conditions for most waterway workers were not as good as those that existed at Honeystreet wharf and at Broad Quay in Bath for example, regular flooding, inadequate accommodation and unhygienic conditions making life very hard for many of them.

In addition to local trading concerns, larger companies such as the engineers and crane makers Stothert and Pitts of Bath, various gas companies and breweries, and the Seend Iron Works, together with other iron and brass founders and other significant manufacturers, all used the Navigation as a means of importing raw materials and delivering finished products to their customers. Some of these organisations contracted out the transport work to local carriers whilst others employed their own fleets of barges. During this period, potential competitors such as the Midland Railway Company, that operated a branch line between Bristol and Bath, used the Navigation to augment

89 *A panoramic view of Honeystreet wharf in around 1910 showing the various buildings. A Kennet barge is moored in front of the boat/barge-building shed, and the timber conversion yard can be seen on the far side of the canal.*

90   An unusual view of Honeystreet wharf showing the launching ways in position in front of the building shed just behind the moored narrow boat. Some of the cottages owned by Robbins, Lane and Pinniger and rented to their employees can be seen on the left of the canal.

91   A drawing of Unity at Honeystreet with a cargo of logs which appear to take up all the space in her hold.

92 *Broad Quay on the River Avon at Bath. This photograph, taken around 1870, shows a number of iron cranes, numerous covered wagons, and two moored narrow boats.*

93 *A 19th-century photograph of Seend iron works. Situated close to the canal, the iron works was in operation from 1856 to around 1888, and used the Navigation to transport its products and probably to obtain certain raw materials used in the iron smelting process.*

94 (Above) A rare photograph of a Kennet barge under tow. Taken from Hungerford road bridge, the image shows the barge heading east away from the bridge, although the horse cannot be so clearly seen as it moves into the shadow of the footbridge.

95 (Right) A 1952 photograph showing a John Krill narrow boat delivering grain at West Mills wharf near Newbury.

96 *Hungerford wharf in 1908 showing a barge crew loading timber.*

its core business and its own fleet of barges and narrow boats supported and extended its rail-based freight and passenger services.

The main cargoes carried on the Kennet and Avon centred on bulk materials, such as coal, timber, stone, bricks, grain and metal ores, and these were often moved in large quantities. Farm and other produce, together with more specialist building materials, chemicals, and numerous items of a more luxurious nature, were increasingly carried to destinations along the entire length of the Navigation. Relatively unusual items occasionally formed part of a vessel's cargo, two examples being stone plinths for statues for a large country estate near Salisbury, and a consignment of 2,000 leeches sold by a London importer of leeches to a client in Bath.

The loading and unloading of cargoes, whether bulk or in smaller quantities, took place at wharves using distinctive wooden, or sometimes cast-iron, hand-operated cranes. Company agents and toll collectors were stationed at several of the Canal Company's own wharves, particularly those that were strategically placed, and the larger wharves often had extensive facilities associated with them such as gauging stations, warehouses, dry docks, and boat and barge basins. In addition to wharves owned by the Canal Company, which in 1823 totalled 19, numerous private wharves, both large and small, existed on the Navigation. Over the period it is likely that at least 90 such wharves were in operation, all of which were owned and/or worked by companies, partnerships or individuals.

97 A rare photograph of barrels of plasticine from William Harbutt's plasticine factory near Bath. The barrels are being loaded by rolling along a plank before stacking in the cargo space of the waiting narrow boat.

98 Two wide boats moored above Newbury lock in around 1900.

KENNET AND AVON CANAL NAVIGATION.

RULES, ORDERS, AND BY-LAWS.

Made and reduced into Writing by the Committee of Management of the Affairs and Business of the Company of Proprietors of the Kennet and Avon Canal Navigation, and having the Common Seal of the said Company thereto affixed, by order of the said Committee, at a General Quarterly Meeting of the said Committee, held on the 26th day of June, in the year of our Lord, 1827, pursuant to the Powers of an Act of Parliament made in the 34th year of the Reign of his late Majesty, King George the Third, intituled "An Act for making a navigable Canal from the River Kennet, at or near "the Town of Newbury, in the County of Berks, to the River Avon, at or near the City "of Bath, and also certain navigable Cuts therein described," and of the several other Acts passed for varying the line of the said Canal and for amending the said Act, some or one of them.

99 *An 1827 copy of the rules, orders, and by-laws of the Canal Company that traders on the Navigation were expected to abide by.*

Although bulk cargoes such as stone, iron and timber were important, coal continued to be the stable trading commodity of the Navigation. Coal from fields in the Bristol, Gloucester, Shropshire, South Wales and Coventry areas was carried, but the opening of the Somerset Coal Canal in 1805, which joined the Kennet and Avon near Limpley Stoke, constituted a major change and meant that coal from the Somerset coalfields could be moved without having to be transported via Bristol. Initially, this meant the supply of coal from other sources suffered, and it wasn't until the towpath on the River Avon was completed in 1812 that coal from the other supply areas was able to compete with that from Somerset. This is evident from an 1812 Canal Company report to the proprietors, which confidently states that as a result of the Avon towpath having recently been constructed:

REPORT

OF THE

Committee of Management

TO THE

COMPANY OF PROPRIETORS

OF THE

KENNET AND AVON CANAL.

IN communicating to the Proprietors a Statement of the Situation of their Affairs, your Committee have, on the present Occasion, subjoined an Abstract of the Whole Receipts and Expenditure, from the commencement of the Undertaking to the 29*th* of May last, conceiving that as the Amount of the Expenditure, referred to in the concluding Part of the Report of last Year, has now been ascertained, it would be satisfactory to the Proprietors to have, in one View, a Statement of their pecuniary Situation.

The Proprietors will observe that the Debts owing from the Company amount to the Sum of £31,942 13 3; to set against which the Debts due to the Company, and the Value of such Lands and other Property as may be converted into Money, without Disadvantage, amount to £23489 0 11. The Balance to the Debit of the Concern is therefore only £8453 12 4. But it appears to your Committee from the Report of the Superintendant, that a further Sum of £23,100 will be wanting for sundry Buildings, Wharfs, Weighing Houses, and other Improvements which it will be expedient to make for the Accommodation of the Trade ; and there is also a Debt of £1666 13 4 on Optional Notes, for which the Company are liable to be called upon. These Sums make together £33,220 5 8, being the Whole Amount necessary to complete the Works and discharge the Debts of the Company; and as the Proprietors have proposed to add this to the Sum intended to be raised under an Act, to be applied for in the next Session of Parliament, to pay for the Kennet Navigation, there will be no Necessity for selling the Wharfs, Warehouses or River Avon Shares, or of reserving the future Income for these purposes.

The 17 River Avon Shares are calculated, in the annexed Statement, at their Cost Price only, but there can be no doubt that they would produce more than double that Amount if offered for Sale.

The Sum of £3353 19 7 expended in obtaining the Towing Path Act will be repaid to the Company, from the Tolls of the Avon Navigation, with Interest at £5 per Cent. but this is not at present valued upon amongst the convertible Assets of the Company, as it will be necessary to advance more Money, on the same Terms, for making the Towing Path, as after mentioned.

The Tonnage for One Year ending May 31st last, amounted to £18,644 1 6.

100   (*Above and Opposite*) *The 1812 Canal Company management report to proprietors outlining the financial and trading position to date, and advising amongst other matters that it is now necessary to construct a towing path on the River Avon section, thus facilitating more effective trading on the Navigation.*

Your Committee have been assured by Traders and others, that these Improvements will occasion a Reduction in the Freight of Goods of at least 2 shillings per ton, they therefore confidently expect that besides the Advantages in the Conveyance of Merchandize, the Inconvenience hitherto felt by having only one Source of Supply of Coals will, in a great degree, be obviated, and Encouragement be afforded to the Proprietors of Collieries in Gloucestershire, Monmouthshire, and at Nailsea, to send that Staple Article of Tonnage along the Kennet and Avon Canal to Devizes, Newbury, Reading, and the intermediate Country.

A year later, however, the Canal Company were reporting that the reason income for the year had not increased was due in part to 'the want of a supply of Coal sufficient for the demand of the trade'. Some years later it reported that, 'In the last Winter there was considerable diminution on the Tonnage of Coals, but the

[2]

When it is considered that the Canal has been open only 18 Months, that the Improvements on the Avon Navigation are not yet effected, and consequently that the Supplies of Coals, Iron, Copper, and other heavy Articles through Bristol and its Neighbourhood, and from South Wales, have not yet been fully facilitated, and that other Causes of a temporary Nature operate at present against the Tonnage of the Canal, your Committee trust that a progressive and considerable Increase in the Tolls may be confidently anticipated.

Another Year's Experience has fully confirmed the Opinion already expressed by your Committee, as to the abundant Supply of Water in all Seasons of the Year, and the most satisfactory Assurances have been given by the Superintendant of the good State of the Works, with a few Exceptions, specified by him, and which are provided for together with the New Works in the before mentioned Sum of £23,100.

The Bath and Bristol Canal Company having suspended for the present the Execution of that Work, it has been deemed advisable immediately to make a Horse Towing Path to the River Avon, under the Powers of an Act passed in the Year 1807, and the Proprietors of that Navigation have resolved on other Improvements.

The Expence of executing this Towing Path as well as of obtaining the Act of Parliament authorizing the same will be repaid to the Company, with Interest, from the Tolls of the Avon Navigation, and the Measure (which was unanimously approved of by the General Meeting of the 1st Instant) is very desirable, even considered as a temporary Accommodation to the Trade, as it will materially facilitate the Passage of Barges, and be a great Improvement on the present Mode of Towing by Men.

Your Committee have been assured by Traders and others, that these Improvements will occasion a Reduction in the Freight of Goods of at least 2s. per Ton, they therefore confidently expect that, besides the Advantages in the Conveyance of Merchandize, the Inconvenience hitherto felt by having only one Source of Supply of Coals will, in a great Degree, be obviated, and Encouragement be afforded to the Proprietors of Collieries in Gloucestershire, Monmouthshire, and at Nailsea, to send that Staple Article of Tonnage along the Kennet and Avon Canal to Devizes, Newbury, Reading, and the intermediate Country.

The Proprietors have already been informed of the important Purchase of the Kennet Navigation, and other Property appurtenant thereto, made by the Committee, under their Authority; for the Purpose of possessing the whole Line of Navigation to Reading.

To enable the Company to make good this Purchase and to discharge their Debts, it has been resolved to apply to Parliament, in the next Session, for Power to raise a sufficient Sum of Money, by the Creation of a further Number of Shares, on the Terms already published, and thereby to secure the Appropriation of the future Income of the Company for a Dividend amongst the Proprietors.

C. DUNDAS,
Chairman.

MARLBOROUGH,
JULY 21, 1812.

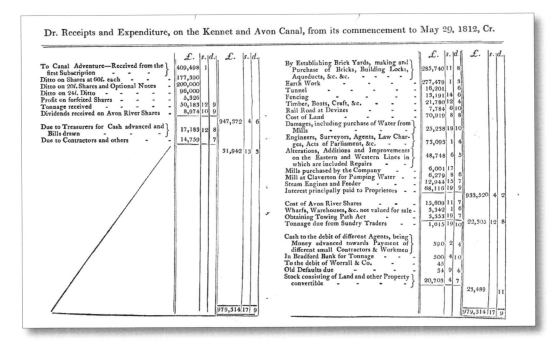

Dr. Receipts and Expenditure, on the Kennet and Avon Canal, from its commencement to May 29, 1812, Cr.

	£.	s.	d.	£.	s.	d.		£.	s.	d.	£.	s.	d.
To Canal Adventure—Received from the first Subscription	409,498	1					By Establishing Brick Yards, making and Purchase of Bricks, Building Locks, Aqueducts, &c. &c.	285,740	11	8			
Ditto on Shares at 60l. each	177,390						Earth Work	277,479	1	3			
Ditto on 20l. Shares and Optional Notes	200,000						Tunnel	16,201		6			
Ditto on 24l. Ditto	96,000						Fencing	13,191	14	6			
Profit on forfeited Shares	5,526						Timber, Boats, Craft, &c.	21,780	12	4			
Tonnage received	50,183	12	9				Rail Road at Devizes	7,784	6	10			
Dividends received on Avon River Shares	8,974	10	9				Cost of Land	70,919	8	8			
				947,372	4	6	Damages, including purchase of Water from Mills	25,238	19	10			
Due to Treasurers for Cash advanced and Bills drawn	17,183	12	8				Engineers, Surveyors, Agents, Law Charges, Acts of Parliament, &c.	73,093	1	4			
Due to Contractors and others	14,759		7				Alterations, Additions and Improvements on the Eastern and Western Lines in which are included Repairs	48,748	6	5			
				31,942	13	3	Mills purchased by the Company	6,001	17				
							Mill at Claverton for Pumping Water	6,279	8	6			
							Steam Engines and Feeder	12,944	15	7			
							Interest principally paid to Proprietors	68,116	19	9	933,520	4	2
							Cost of Avon River Shares	15,609	11	7			
							Wharfs, Warehouses, &c. not valued for sale	3,342	1	6			
							Obtaining Towing Path Act	3,353	19	7			
							Tonnage due from Sundry Traders	1,615	19	10	22,305	12	8
							Cash to the debit of different Agents, being Money advanced towards Payment of different small Contractors & Workmen	590	2	4			
							In Bradford Bank for Tonnage	500	4	10			
							To the debit of Worrall & Co.	45					
							Old Defaults due	34	9	4			
							Stock consisting of Land and other Property convertible	20,703	4	7			
											23,489		11
				979,314	17	9					979,314	17	9

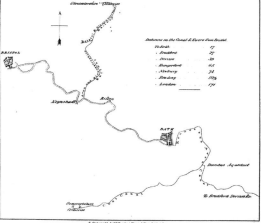

101 *An undated Canal Company management report informing the proprietors of the death of Charles Dundas, declaring a dividend, and advising of the arrangements being put in place to ensure that the coal trade on the Navigation is enhanced by bringing coal to the canal from the Gloucestershire collieries.*

increase in the general Trade on the Canal more than covered this deficiency.'

Although a great deal of coal was transported on the Kennet and Avon, the level of trade rarely seems to meet the Canal Company's expectations, despite the fact that the building of the company-owned Avon and

Gloucester horse railway, which terminated at Londonderry wharf on the River Avon, did allow coal from the collieries in Gloucestershire to be brought more readily to the Navigation. Later, under the influence of increasing railway competition, the volume of coal and associated company receipts began to decline severely.

7 | THE END OF AN ERA:
RAILWAY COMPETITION, DECLINE IN TRADE, CHANGED OWNERS

Horse railways, such as the temporary one established alongside the canal at Caen Hill and the more permanent arrangement at Londonderry wharf on the River Avon, were sometimes built to assist trading activities that took place on rivers and canals. Locomotive railways, however, were very different, and as they spread across the country they began to affect the near monopoly position enjoyed by the waterway carriers. As Charles Hadfield notes in *British Canals*, towards the end of the first quarter of the 19th century, when the first public railway to be successfully operated by locomotive engines was opened between Stockton and Darlington, a tremendous outburst of speculative railway building commenced. In a similar fashion to that required for canal schemes, permission via Acts of Parliament were necessary before railways could be constructed. It was during these early years of the century that the Great Western and other major rail initiatives of the next decade were first projected.

Writing about those early developments, W.T. Jackman points out in *Transportation in Modern England* that once London had been connected by railway with centres in the Midlands and the north west of England, attention became firmly focused on securing a connection between the capital and Bristol. Although Bristol's status as as the country's

second largest port had been eclipsed by Liverpool by this time, it was still a major trading centre. Agitation for a rail connection between Bristol and London had been taking place since as early as 1823. In *The Kennet and Avon Canal*, Clew mentions that sufficient interest in a London-Bristol rail link had been generated by 1833 for the eminent engineer Isambard Kingdom Brunel to be engaged to survey possible routes. Once this work had been completed and a favoured route agreed, a Bill was submitted to Parliament seeking powers to construct railway connections from London to Reading and from Bath to Bristol, with approval for the intermediate section to be sought the following year.

The Bill for what was to become the Great Western Railway was supported by many of the towns along its proposed route. Some landowners, however, as well as those who still viewed rail transport with suspicion, still considered the Kennet and Avon waterway to be the most viable and cost-effective means of transporting goods in the area and opposed attempts to enact the new legislation. The Bill survived all objections in the House of Commons, but was rejected during its second reading in the House of Lords. A second attempt by the Bill's supporters was then swiftly made, and this time they overcame all opposition. Legislation was subsequently enacted in the form of the Great Western Railway Act of 1835.

In *Inland Waterways of England*, L.T.C. Rolt has indicated that the downfall of the canal industry in the face of railway competition was as swift as its meteoric rise to prosperity. Railways unquestionably held many advantages, especially for the convenient and practical transport of passengers, but for the bulk carriage of heavy goods, water transport still provided a viable alternative. Unfortunately, canal companies had exploited their monopoly position for so long, and had in many cases lost public sympathy and support; consequently their continued role in enabling bulk carriage to take place was never fully appreciated. A further important disadvantage was that canal companies were rarely waterway carriers but merely toll takers, and the tolls on individual canals within the inland waterway network often differed widely. As a consequence, small traders who operated their narrow boats on more than one of the interlinked waterways found that expeditious through-traffic working and the quotation of through rates to potential customers was nearly impossible.

Railways and canals were also widely dissimilar in nature and, as Jackson relates, the general perception of the railway was one of vitality, energy and efficiency; the large trains, their prompt arrival and departure, and the speed and bulk of the engines were all subjects of admiration, and stood in stark contrast to the quiet, unseen canal and its slowly plodding barges. Those who supported the 'old technology' were increasingly marginalised, and water transport was rarely perceived as an effective alternative to the rapidly expanding railways.

Despite the growing threat posed by railway competition, trade on the Kennet and Avon continued. Although toll and tonnage receipts had gradually fallen, from about 1835 they began to increase significantly, reaching a peak in 1840 after which time they began to decrease

again. Ironically, much of this increase was due to traders on the Navigation transporting building and other materials used to construct the new railway. John Rennie, the Navigation's creator, John Ward, its first Principal Clerk, and Charles Dundas, the Canal Company's Chairman, who had succumbed to cholera in his 81st year, had all passed away by this time. The management had passed on to others, and none of the original champions of the once great Kennet and Avon trading route lived to witness the onset of its decline.

In general, canal companies facing railway competition tended to go on the defensive and attempt to protect their rights rather than seek to secure more trade. The Kennet and Avon Canal Company initially pursued this course of action, although it later attempted to compete with the railway. In 1837 a decision was taken to allow carriers to use the Navigation by day and night in a bid to increase traffic. In this instance, the resulting activity created a water shortage on the canal section, which in turn prompted the imposition of temporary restrictions on the draught of boats and barges. Unfortunately, these restrictions led to a decrease rather than an increase in tonnage carried, not at all what the Canal Company had intended.

Improvements to passenger services were also made, with Scotch boats taking pride of place. Clew notes that such vessels decreased passage times, and on the service between Bradford-on-Avon and Bath, for example, an average of some 40 passengers were carried in each direction on a daily basis. The Canal Company introduced new regulations for the fly-boat service between Bath and Reading, which encouraged the wider use of that service.

The section of the Great Western Railway between Bristol and Bath was opened in 1840, and a year later the complete line between Bristol and London became operational. The immediate effect was that almost all waterway-based traffic between the two cities was lost to

102 *A photograph showing the close proximity of the Great Western Railway and the Navigation; a frequent occurrence. The width of a lane separates the two before it disappears over a hump-backed canal bridge.*

the new railway, although local traffic continued at normal levels. Later, even local traffic was affected, with receipts for the period 1840-1 dropping from just over £51,000 to around £40,000 the following year.

In an attempt to reclaim its trade levels, the Canal Company reduced its tolls by 25 per cent, as well as reducing the dividends it expected to pay to shareholders by 50 per cent. Committee expenses and employee wages were cut at the same time, the number of permanent staff employed was reduced from 122 to 100, and the level of repair work was also reduced. As a consequence, the committee were able to report that:

> The net Tonnage on the Canal and Rivers for the past year, is less than the preceding year, but It must be observed that the great reduction in the rates of Tonnage have been in operation during the whole year, and considering the extreme mildness of the season, Railway competition, and the continued severe depression of Trade, the result has been more favourable than was anticipated. Your committee have as in

the preceding years had the consideration of expenses constantly in view, and have satisfaction in calling attention to the reductions under the Heads of Repairs, Salaries, Committee expenses, etc. amounting to the sum of £3,500.

These actions resulted in a slight increase in tonnage carried in the following two years, and prompted the introduction of an additional fly-boat service between Bristol and London. In practice, however, this later initiative was unsuccessful and the service was discontinued the following year. Around this time, modifications at both Claverton and Crofton pumping stations were made to ensure that both plants became more efficient. Further reductions in tolls took place in 1845, and although this tended to increase the tonnage carried, the company's income continued to fall.

Other attempts were made to reduce costs associated with the canal and find some way of competing successfully with the railway. One rather drastic course of action was included in the 1845 shareholders' report:

The Committee have under their consideration the expediency of converting the canal into a Railway, and they intend taking immediate steps to obtain every information and order an actual Survey to be made, that they may be prepared (if it be found practical) to lay before the Proprietors the best mode of proceeding.

An engineer was appointed to investigate this proposal. Clew notes that in a report prepared later that year the engineer recommended the canal be retained, but that a railway be constructed alongside it, to be called the 'London, Newbury, and Bath Direct Railway'. There was some support for this proposed railway, that would presumably have enabled the Canal Company to run rail and passenger services in conjunction with those that were waterway based, but the associated Parliamentary Bill was, after its second reading, referred to a Select Committee. The Select Committee, set up to consider a number of railway related Bills placed before Parliament, finally ruled that the Canal Company's proposed Bill should not be allowed to continue to the next stage and that it would be more appropriate if the Kennet and Avon operations were taken over by the Great Western Railway.

Despite this suggestion, the drive for economy continued, with lock-keeping, canal maintenance, and repair work being put out to contract. Over the next two years, as a result, the Canal Company reported that much of the lost traffic had been recovered. At around this time it was decided that the Company should expand its own carrying trade, and 32 vessels, including Severn trows and Kennet barges, together with a number of horses, were subsequently acquired. Along with the lease of several extra wharves, the cost of this initiative had reached £12,000 by 1849 according to Clew. An extra daily barge service between Bristol and London with a journey time of five days

was now on offer, but the new arrangements were unsuccessful and were suspended the following year.

Upward trends in traffic volume were not matched by increased income from toll receipts, and price-cutting by the railway company tended to make matters even worse. It was also evident that Navigation tolls could not be reduced further if the Canal Company was to remain solvent. Clew recounts that the last dividend was paid in 1850 and there seemed little hope of any more. The Company now had one option left and in 1851 it offered the waterway to the Great Western Railway. Later that year an agreement was reached whereby the railway company would take over all assets and stocks, together with debts, liabilities and contracts, and would maintain the Navigation and associated works in a condition suitable for its continuing use. An annual payment of 6s. per share was to be paid in perpetuity to all Kennet and Avon shareholders and the Canal Company was to continue administering the undertaking on the railway company's behalf until such time that associated legislation could be enacted. The final transfer of the Navigation was subsequently authorised by the Great Western Railway Act No. 1, which received Royal Assent in 1852.

Charles Hadfield has intimated that the capitalised cost to the Great Western Railway of taking on the waterway was nearly £211,000, although the original cost of constructing the canal had been closer to a £1 million. By the end of the 1840s, railway companies owned nearly one fifth of the canals of Great Britain, all normally acquired for what were considered good commercial reasons. At the point of takeover, however, the Kennet and Avon was not perceived as a competitor, so why was the Great Western Railway Company so interested? The overriding reason, as Clive Hackford points out in his article 'G.W.R. – Saviour or Destroyer', must have been that the Kennet and Avon Canal Company was not a viable rival as it stood but its continued existence posed a threat inasmuch as it could be converted into a competing railway in the future. It was a risk the Great Western was not prepared to take.

At first the Great Western tried to run the Navigation as a profitable adjunct to its rail operations, and this is evident from a report to its own shareholders prepared some 12 months after the Kennet and Avon had been acquired:

> There is every reason to hope that the arrangements for utilising the usefulness of that undertaking – when worked in unison with the railway – will prove to all parties interested, the advantages as well as the prudence of these arrangements.

The development of integrated transport arrangements appears to have been the railway company's initial intention, with high volume, low value goods, where delivery speed was not so important, being transported on the Navigation, whilst passengers and more valuable cargoes would be carried on the railway. Within the space of 10 years, however, this policy changed and increasing losses on the waterway meant that most effort was concentrated on rail operations. When waterway maintenance contracts came up for renewal, the final terms of the policy tended to place contractors at a financial disadvantage. As a consequence, a general decrease in the standard of maintenance on the Navigation became increasingly evident, and some important operations such as ice-breaking were even stopped for a period.

103 *An excursion party at Bradford-on-Avon wharf. The fine display of Union Jacks suggests that the group might well be celebrating Queen Victoria's Diamond Jubilee in 1897. The narrow boat has been splendidly painted and bears little resemblance to the drab working boats normally used on the Navigation.*

GREAT WESTERN RAILWAY.

KENNET & AVON CANAL.

NOTICE OF STOPPAGE !

NOTICE IS HEREBY given that the **Water** will be withdrawn from the Canal at the

Cobbler's Lock

in HUNGERFORD MARSH, on Monday, the 9th and Tuesday the 10th of September next, (both days inclusive) for Repairs to the Lock, during which time the Traffic through the Lock will be suspended.

CHARLES F. HART,
Canal Engineer.

Engineer's Office, Devizes,
28th August, 1889.

104 *Notice of maintenance work to be carried out by the railway company, dated 1889.*

In apparent disregard for the spirit of the 1852 Act and of a further piece of legislation, the Regulation of Railways Act 1873, which required the Navigation to be maintained in a usable condition, various means were enlisted by the Great Western to discourage trade on the Kennet and Avon. These included regularly increasing tolls, closure of certain wharves, staff reductions, and attempts to coerce traders into sending goods by rail instead. Complaints from and disputes with carriers concerning tolls, maintenance and water shortages were frequent occurrences. One waterway partnership, Gerrish and Sainsbury, refused to submit to restrictions on certain locks, and on two separate occasions broke open the secured lock gates so that their boats could pass through. Although some damage resulted, the railway company was unable to take action as the detention of trading vessels was legally unenforceable.

Further disputes over a number of issues, including excessive charging, eventually led to Gerrish and Sainsbury seeking redress in the courts. Their action resulted in the Great Western being ordered to pay compensation for trying to restrict free passage of the boats in question.

Despite the court's ruling, indifferent maintenance continued and this, together with continually increasing tolls, deterred many carriers from using the waterway. The gradual working out of the Somerset coalfields also had an effect, with traffic along the coal canal to its juncture with the Kennet and Avon at Dundas wharf having completely stopped by the end of the 19th century. A similar fate befell the Wilts and Berks, the second feeder canal, upon which all trade ceased by 1900. By this time the Kennet and Avon was regularly recording a loss each year, and the increased tolls that in 1906 were said to be higher than on any comparable waterway, had succeeded in driving away almost all through-traffic regardless of type or quantity of cargo carried.

General economic decline during the First World War led to further reductions in the traffic, although the Navigation was used for a time by the Admiralty to move a number of small inshore naval craft from London to Bristol, and troops were trained on the canal at Devizes so that they could operate boats on the canals of Belgium. Some trading on the Navigation continued after the war, although in 1920 the railway company further increased tolls, this time by a staggering 150 per cent. Six years later the company announced its intention to apply to the Ministry of Transport for an order to close the waterway. Such was the opposition to this proposal that by 1928 the plans were abandoned, at least for the time being. Instead, in an unprecedented move that may well have resulted from this opposition, as well as pressure from traders such as Robbins, Lane and Pinniger to honour earlier agreements,

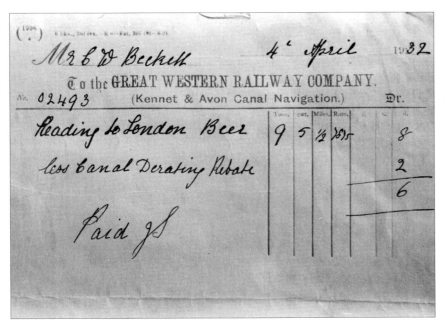

105 *A Great Western Railway receipt for carriage on the canal of nine tons of beer, dated 1932.*

106 Great Western Railway toll rates in 1868. Foodstuffs and luxury items attracted the highest tolls, whilst the cost per ton for bulk carriage could be significantly less.

107   *A popular event on the canal near Devizes, possibly a Sunday school outing aboard a narrow boat, c.1900.*

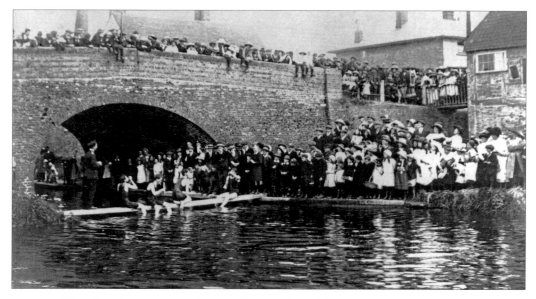

108   *Although trade activity was declining, the annual canal-based swimming race for cigarette prizes was a popular event for the residents of Hungerford, c.1913.*

109 *This narrow boat was owned and operated by the Midland Railway Company. Leaving Dundas wharf for Bath, the boat is providing a feeder service for passengers who may wish to travel from Bath on the company's rail service to the Midlands, c.1900.*

the railway company embarked on an extensive programme of dredging, bank maintenance, and lock gate replacement.

Irrespective of this, the Great Western continued to discourage trade on the Navigation, regardless of the fact that many traders still wanted to use it. In addition, wharves and associated buildings were sometimes sold off or reduced in size. The once extensive wharves and company buildings at Newbury, which had been progressively reduced as trade diminished, were sold to Newbury Corporation in 1931 and eventually became a vehicle park. Over the next 10 years the railway company fulfilled

its minimum statutory obligations by keeping the Kennet and Avon navigable, but its general approach meant that trade continued to decline, so that trading activity had fallen to an all-time low by the beginning of the Second World War in 1939.

Very little movement took place on the waterway during the Second World War, apart from a flurry of activity when an invasion by German forces was considered likely. The canal was seen as a natural barrier to this, and concrete pill boxes were gradually constructed along parts of its length in an attempt to increase its defensive capability. The materials required for

110 *During the First World War, soldiers of the Transport Division were trained in boat-handling skills at Devizes. This was considered necessary to equip them for war duties on the canals of Europe with the British Expeditionary Forces.*

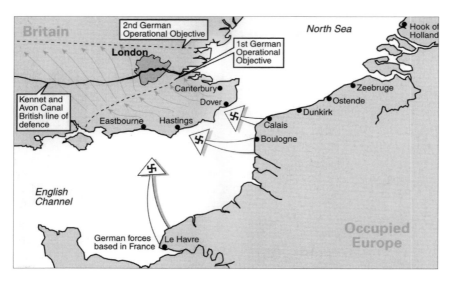

111   A map showing the strategic role planned for the Kennet and Avon Canal as a primary line of defence in the event of an invasion by German forces during the Second World War.

112   The defensive line of the Kennet and Avon Canal was enhanced by the construction of brick and concrete gun emplacements known as 'Pill Boxes.' Although most could easily be seen, a few, like this one at Honeystreet, were well disguised.

113   This photograph shows the Devizes fire brigade using the canal to carry out a pump test during the Second World War. In line with the canal's defensive role, concrete anti-tank barriers have been erected at the bridge entrance.

114 *Bomb damage at Locksbrook Bridge in Bath where it crosses the Navigation, c.1942.*

these structures were naturally transported on the Navigation, and there is some evidence to suggest that the water level in parts of the canal section was lowered to prevent flooding in the event of bomb or other associated damage.

During the war the railway company considerably reduced its maintenance programmes, and although no action was actually taken, the possibility of closing sections of the canal was still under consideration. Lack of use took its toll on the waterway, and low water levels and partial blockage by weed made through passage very difficult after the war. It was not impossible, however, as was demonstrated by the journey of Lord Bingham and George Day in 1947 on the narrow boat

Hesperus, which towed a landing craft called *Tranquil* for delivery to a Newbury client.

By 1948, when railway nationalisation took place and the newly appointed Railway Executive assumed administration of the Navigation, the Kennet and Avon was still just about navigable. However, even the few committed local traders who had been brought up on the Navigation found using it difficult, time-consuming, and frustrating in the extreme. Travel on the Navigation now required a great deal of additional effort on the part of carriers, who often needed assistance from maintenance gangs to clear a passage. This assistance was provided by the Railway Executive as a cheaper alternative to normal waterway maintenance, although, as Clew points out, it was only freely available during normal working hours on Monday to Saturday mornings. At all other times the carriers had to pay for the service at the requisite overtime rates, and this could be expensive as gangs of up to 12 men were typically involved.

The overall administration of Britain's waterway system was passed in turn to numerous Government departments, whilst maintenance boats on the Kennet and Avon were allowed to rot at their moorings. As future prospects appeared bleak, canal trader John Gould, and later the Willow Wren Carrying Company Ltd, took legal action

115 *A Great Western Railway spoon dredger being towed near Limpley Stoke, c.1920.*

116 *Taken near Reading, this photograph shows a steel mud boat used to hold the waste from a dredging operation.*

117 *The narrow boat, Columba, owned by John Krill, shown at Newbury's West Mills wharf after carrying salt from Cheshire in 1950.*

against the administering department, the Docks and Inland Waterways Executive of the British Transport Commission, for not carrying out their statutory obligations. Before the cases came to court, however, the British Transport Commission published the findings of a 1955 review which recommended that they should be relieved of their responsibilities for deteriorating and uneconomically viable canals such as the Kennet and Avon. Later that year the Transport Commission announced that in view of these findings it intended to abandon the complete canal section between Reading and Bath, and they would be placing an associated Bill before Parliament with a view to initiating closure.

8 | PLUCKED FROM OBLIVION:
SAVING THE WATERWAY

Lack of maintenance and repair ensured an inevitable deterioration of the waterway. Despite continuing tar barge and other traffic on the River Avon, infrequent journeys by mainly Newbury carriers, occasional trips by adventurous pleasure craft, and the annual Devizes to Westminster canoe race, the River Kennet and the canal sections of the Navigation in particular now appeared to be doomed. But powerful voices against closure were beginning to be heard. As Clew relates, the Transport Commission proposals resulted in protests from many quarters, including county councils, borough councils, rural and parish councils, and individuals living along the whole length of the canal.

The leisure potential for inland waterways was increasingly becoming an issue, and interest groups were beginning to speak out against planned waterway closures in many different parts of the country. One of these groups, the Inland Waterways Association, formed some ten years earlier and working through a network of branches, conducted vigorous campaigns for the retention and continuing use of waterways and encouraged their use for leisure purposes. An early cause to be fought by the Association was the proposed abandonment of the Kennet and Avon, and a local Inland Waterways Association branch was duly formed to assist the process. This in turn led to the formation

of a Kennet and Avon Canal Association; the forerunner of the present day Kennet and Avon Canal Trust.

Interest and enthusiasm for averting closure rapidly grew, and what Clew terms a 'fighting fund' was established, a petition organised, and protests planned. Attempts to arrange an Easter protest cruise on the canal had to be abandoned, however, when certain locks were inspected and declared unsafe by the divisional engineer of British Transport Waterways, the body formed to administer the country's inland waterways when the Docks and Inland Waterways Executive was separated at the beginning of 1955. Undaunted by such setbacks, the Kennet and Avon Canal Association organised protest meetings in each of the main towns along the canal route, with copies of the petition being distributed to other towns and villages in the vicinity. Only those living near the waterway were allowed to sign the petition but, even so, more than 22,000 people responded, and the associated sheets of signatures were bound into two large, loose-leaf volumes.

The completed petition was to be delivered to the Queen, and it was planned it should travel the whole distance from Bristol to London by water. As Peter Lindley Jones recounts in *Restoring the Kennet and Avon Canal*, the petition left Bristol on 4 January 1956 and was taken by cabin cruiser to Bath,

118   *A poster advertising a 1955 protest meeting at Reading Town Hall.*

and thence by canoe to Thames Ditton, where it was loaded, canoe and all, onto a river cruiser for transportation down the Thames to Westminster. It was a journey of 157 miles that, allowing for publicity and other stops en route, took 12 days to complete. Amidst extensive press, film, radio and television coverage, the canoe, with petition safely stowed inside, was then carried through the streets of London to the Ministry of Transport in Berkeley Square, with a request that the document be forwarded to the Queen forthwith.

The legal action by John Gould referred to earlier, together with an additional application for an interim injunction to prevent further deterioration of parts of the canal section, was heard late in 1955. The judge did not uphold the injunction, although he did criticise the British Transport Commission's attitude, and they eventually admitted to being in default of their statutory obligation to keep the waterway navigable. Damages of £5,000 were awarded to John Gould but, and having agreed that it would be up to Parliament to decide the future of the Navigation, the judge made an order that no further proceedings should take place until the pending Parliamentary Bill had been considered. In view of this ruling, the Willow Wren action was dropped and a settlement for damages made out of court.

119 (Above) The petition against closure of the waterway on its way to Westminster in 1956 for eventual delivery to the Queen.

120 (Right) The motor cruiser rally at Reading in 1956 in support of the movement to save the waterway.

121   *A horse-drawn trip boat near Kintbury, operating as part of the fundraising initiatives.*

A British Transport Commission Bill was published at the end of 1955 seeking authority for a scheme whereby the rights of Navigation between Reading and Bath, the canal section of the waterway, would cease. The British Transport Commission also requested a release from any obligation to keep the canal navigable during the estimated five years the scheme would take to prepare and implement. At the same time the Government, recognising the growing and widespread concerns that existed for the future of the Kennet and Avon and for inland waterways in general, had set up a Committee of Inquiry. The Bowes Committee was responsible for examining the future of the whole inland waterways system.

The Bill was first placed before Parliament in the spring of 1956, with a Select Committee

subsequently being established to hear arguments for and against its proposals. This group agreed to support the British Transport Commission's case, but stipulated that the obligation to keep the canal open should remain until 1960 and maintenance should continue to be carried out during the intervening period. On passing its third reading, the Bill was put before the House of Lords, and although further opposition resulted in some clause changes, it was duly placed on the statute book to become the British Transport Commission (No. 2) Act of 1956.

Some 14 months later, in the summer of 1958, the findings of the Bowes Committee were published. The Committee recognised that a small number of waterways could still be utilised by commercial traffic, whilst the vast majority were worth retaining and keeping navigable for the use of pleasure craft. Bowes Committee members also decided that each waterway should be considered on its own merits in order to establish how best it could be used, and the co-operation and financial involvement of interested parties should be sought in all associated redevelopment schemes. Significantly, as Lindley Jones points out, the Bowes Committee had broken away from the concept of abandonment of the canal, and was now suggesting a new and positive approach that involved redevelopment of existing facilities. In a White Paper issued in 1959 the Government concurred with these recommendations, proposing that they be implemented as an interim policy to be tested on an experimental basis over the next two years.

The Bowes Committee also recommended the formation of an Inland Waterways Redevelopment Advisory Committee, to assist in promoting schemes and to make recommendations to Ministers. This arrangement also won support from the Government, and Clew notes that at its inaugural meeting the Advisory Committee issued a list of priorities for action, with the Kennet and Avon foremost amongst these. Meanwhile, in the spirit of the Bowes report, the Kennet and Avon Canal Association had prepared its own redevelopment scheme which it published in 1961, submitting copies to both the Redevelopment Advisory Committee and to Members of Parliament. Prior to this, some lock and other essential repair work had been carried out by British Transport Waterways and volunteers from the Kennet and Avon Canal Association, and the success of this joint effort led to talks between the Association and the British Transport Commission. The talks resulted in a promise from the Commission's General Manager that, within the constraints of funds and manpower availability, he would support the association in its work.

The waterway and its uncertain future was kept on the agenda by constant pressure from supporters, which included tours of inspection organised initially by the Kennet and Avon Canal Association, and later by the Redevelopment Advisory Committee. Such arrangements allowed Members of Parliament and other interested parties to view the canal and its problems at first hand and be shown how action on their part could help provide positive solutions. The lobbying had an effect for, although still undecided on a longer term course of action, the Government extended the interim period of obligation to maintain the waterway in a navigable condition up to the end of 1963. Unfortunately, decreasing use of the waterway by the few remaining carriers, who appear to have succumbed to the lure of the railway, together with a reluctance by all but the hardiest 'boater' to fight his way through, meant that the canal section's general condition was still deteriorating.

Towards the end of 1961 the Redevelopment Advisory Committee made recommendations to the Minister of Transport on the future of the Kennet and Avon, and these were made public in the spring of 1962. Essentially, it was proposed that a joint undertaking of waterways staff and representatives of canal associations should redevelop the waterway. The basic funds allocated for redevelopment would be the annual £40,000 currently assigned to annual maintenance, divided in such a way that half would be used to maintain the canal in its present condition, the other half used for rebuilding work. The programme would involve starting work at both ends of the canal simultaneously and progressing in stages towards the centre. It was expected that volunteer labour would be used extensively to augment paid resources, and that further voluntary effort on a vast scale would be needed to raise the additional funds likely to be required. In view of the enhanced role and

122   *An abandoned narrow boat near Bathampton, c.1935.*

123   *An abandoned maintenance boat near Crofton, seemingly voyaging in a sea of grass, c.1955.*

responsibilities they would have to assume if these proposals were accepted, the Kennet and Avon Canal Association reformed itself into a non-profit company, the Kennet and Avon Canal Trust Ltd, which was later awarded charitable status.

The proposed approach seemed sensible to all concerned, despite the fact the plans could not be implemented immediately. The Government's full support was not yet secured, and the extended period of temporary standstill was not scheduled to expire until the end of 1963. Clew notes that many well-wishers at 'Save the Canal' meetings in Reading and Bath passed resolutions calling upon the Minister of Transport to take notice of his advisory body's recommendations and allow the required work to commence forthwith.

The Transport Act of 1962 provided some peripheral help by separating British Transport Commission interests and establishing a new independent authority known as the British Waterways Board to administer all waterways matters. It was expected that a decision on the Kennet and Avon Navigation's future was likely to be deferred until 1963, when the new waterways authority was to commence operation, but the Government extended the moratorium again, this time until the end of 1967, most likely because a General Election was looming.

The British Waterways Board initiated a policy of constructive co-operation with canal users and associated organisations from the start, and great hopes were entertained for the future. Towards the end of 1963 it issued an interim report on Britain's waterways in which the Kennet and Avon was highlighted as a special case, on which they would be prepared to work with the Canal Trust and other enthusiasts to bring about the required restoration. It also recognised that additional funds on top of those indicated by the Redevelopment Committee were likely to be required, but as long as these could be kept to a modest level was prepared to provide some financial assistance.

Restoration work could not start until the Minister of Transport signified approval of the British Waterways Board's proposals, but a joint decision was made to initiate some major lock repairs on the River Kennet section using volunteers under professional supervision. Work on a drained length of the canal was also carried out to prevent any further leakage. Once this had been completed, efforts were concentrated on the flight of locks near Bath, whilst other schemes were initiated to alleviate landslip problems. During this period the Canal Trust was able to raise

additional funds by charging winter mooring fees for boats moored near Bath. A junior division of the Canal Trust was also formed to interest young people in the waterway. The adventure training allowed much useful canal clearance and other work to be carried out by the individuals concerned.

A comprehensive survey of canals carried out by the British Waterways Board and published late in 1966 showed that many of the assumptions about canals upon which decision-making had been based were incorrect. It highlighted greatly inflated asset values, the fact that abandonment of a canal did not mean expenditure on it could stop completely, and a reminder that sections of navigations such as the Kennet and Avon were in fact rivers and could not be eliminated even if the canal section was. Clearly influenced by the results of this survey, as well as British Waterways Board requests that restoration work on the Kennet and Avon should commence, the Government published a 1967 transport policy White Paper. This accepted the broad

124 *The pleasure boat about to enter this lock will have to move through a sea of weed.*

125 *The weed cutter, Moonraker, designed and built by Canal Trust members. This unusual craft was highly effective in use, c.1967.*

126 *The Youth Division of the Kennet and Avon Canal Trust was formed to harness the energy of the younger generations. Members are shown helping in the restoration of a lock.*

factual analysis provided, and indicated an intention to initiate discussions with interested parties, both locally and nationally, so that it could establish which inland waterways should be kept open for boating and other amenity purposes. Any waterway still considered to have real commercial value would be recognised as being part of a group separate from the amenity network, which would still need to be administered by the Waterways Board.

Later the same year the new Minister of Transport issued a public invitation to all parties interested in the future of Britain's inland waterways, asking for their views on the matter. However, another year was to pass before another White Paper accepted that the nation's existing waterway network should be retained, and that provision should be made for its use for both commercial and leisure purposes. By then, of course, all remaining commercial traffic on the Kennet and Avon had ceased and it was difficult to imagine a case being made for it starting again.

The White Paper stipulated that the British Waterways Board would be given responsibility for maintaining the inland waterways network to a standard that would allow use by powered pleasure craft, and that this new recreational purpose would be recognised for the first time by an Act of Parliament. It encouraged volunteer input to assist in restoration, but awarded cruise-way status only to those waterways it would directly maintain via the Waterways Board. Only parts of the Kennet and Avon had been given cruise-way status and maintenance of the rest would now have to be funded separately.

A new consulting body, the Inland Waterways Advisory Council, was also established. Later to become known as the Inland Waterways Amenity Advisory Council, it had the task of advising the Minister of Transport on any proposals that might affect the waterway

127 *The Canal Trust's 70-ton steam dredger,* Perseverance, *at work during the waterways restoration.*

system. After a period of consultation, the Government's transport intentions as outlined in the White Paper were enshrined in the Transport Act of 1968, and this was to guide policy until such time as new legislation might be introduced.

128 *In this photograph the canal is shown full of vegetation above which broken lock gates rise starkly.*

129   *A deserted lock, its broken gates emphasising the growing dereliction.*

130   *(Below) The gates have disappeared from this lock and the chamber is fast reverting to nature.*

9 | A Phoenix Rising:
Restoration and Renewal

One of the first matters to be considered by the Inland Waterways Amenity Advisory Council was the pilot restoration scheme previously prepared by the Kennet and Avon Canal Trust. Clew notes that the scheme was intended to link cruise-ways at Reading and Newbury, extend the River Avon cruise-way eastwards as far as Dundas aqueduct, and dredge the 15-mile pound between Devizes and Wootton Rivers, all as the first stage of a 10 year restoration programme. Advisory Council members inspected the canal over a three-day period in the summer of 1968, they gave their support and advised the Minister accordingly.

Meanwhile, volunteers continued with lock repairs and involved themselves in major restoration work at Crofton pumping station which, following a nationwide appeal for funds, was now owned by the Canal Trust. Survey and initial restoration

work at Claverton, carried out by engineering students from Bath University, led the way for volunteers to bring the pumps back into operation. It wasn't until 1975, however, that both the Crofton steam engines and the water-powered pumps at Claverton were fully operational again.

In addition to the physical input provided by volunteers in their roles as 'new navvies', working with Waterways Board employees, the Canal Trust and its members augmented budget allocations by raising funds for much of the repair work. This was achieved by donations from within the ranks of trust members and by

131 *A deserted lock, beginning to be reclaimed by nature.*

132 *The remains of a turf-sided lock with Second World War pill box gun emplacement alongside.*

133 *The entrance to a derelict turf-sided lock on the River Kennet.*

arranging rallies and other events. As the pace of restoration quickened, and the requirement for campaigning and raising awareness conversely reduced, more effort was required to find benefactors and donors that could provide the much larger levels of funding that would be required in the future.

The numerous county, district and other local authorities that existed along the Kennet and Avon's route had been very supportive of the 'great endeavour' in the past, and it was only natural that requests for monetary donations would be directed towards them. At the same time a nationwide appeal for funds was set up and, although some set-backs occurred, within two years more than £30,000 had been received from local authorities, with other donors providing more than £50,000. Additionally, as Lindley Jones points out, the steadily increasing trust membership, averaging around fifty new members per month, also augmented income. Trust funds were boosted by admission receipts from the recently opened Crofton pumping station, which at weekends attracted anything up to 3,000 visitors to see the steam engines and pumps being operated.

Government plans to create regional water authorities to take over responsibility from the British Waterways Board were abandoned in 1973 to the relief of many of those affected, including the British Waterways Board and the Kennet and Avon Canal Trust. The aborted plans prompted the Canal Trust to urge the Waterways Board to accelerate restoration work for 1972 and 1973. The ability to do so, however, was dependent on the Board's financial commitments and the capacity of its labour force. Clew records that during 1972 these two things were finely balanced, so the only way additional work could be done was through the use of private contractors, although it was understood costs might increase as a result.

Although restoration work continued, the rising costs of contract working and the high inflation that a downturn in the economy had created were having an adverse affect. The effects of inflation were particularly disruptive, and inflation-generated costs were not always passed on in practice, but by 1975 the Waterways Board had to insist that all outstanding orders should be provided with upgraded estimates before work could commence. Clew notes that at this time around 41 miles of the waterway had been made navigable and 27 locks reopened. It was now evident, however, that extensive additional funds would be necessary if complete restoration was ever to be achieved. For the Canal Trust, stopping or even reducing work activity was never a realistic option, and instead of accepting a gradual slow down it launched a further national appeal with a view to obtaining £500,000, to continue and, if possible, speed up the ongoing restoration process.

It was hoped to generate interest from commerce and industry, but continuing recession meant that funds from these sources were lower than expected, with just over £20,000 being obtained by the spring of 1976. National appeal activity continued, and high profile visits by individuals such as the Duke of Edinburgh were arranged, but the decision was also taken to target grant-awarding trusts and similar bodies in the hope of increasing funding levels. A further Government White Paper of 1977 proposed that the duties and responsibilities of the British Waterways Board should be transferred to a new national water authority, prompting fears that restoration aspirations might be prejudiced, so numerous objections to these proposals were submitted. It wasn't until the results of an investigation into the Waterways Board's operating and maintenance costs had been completed and the recommendations of a Select Committee taken into account that the threat was removed. In 1979 the Government agreed that the British Waterways Board should continue as an independent organisation.

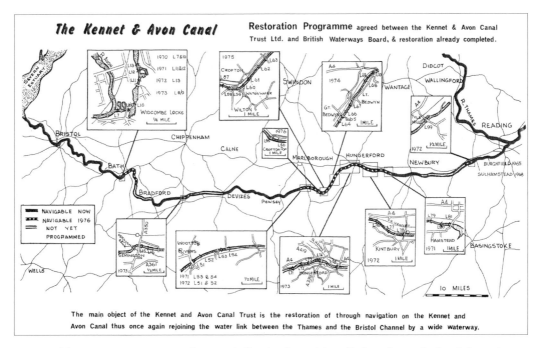

134   *The restoration programme for 1972 indicating the position of locks to be repaired and the sections of the waterway scheduled for restoration by a planned date of 1976.*

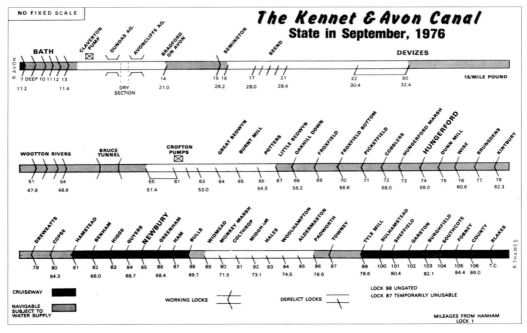

135   *The restoration programme for 1976 in more schematic form than the 1972 programme, indicating working and derelict locks and navigable sections at that date.*

The Transport Act of 1968 had divided the Kennet and Avon into three cruise-ways separated by so-called 'remainder lengths', parts of which were in practice quite navigable. The restoration work carried out up to this point had effectively extended the cruising lengths into the remainder, although many obstacles to full navigation still remained. Job creation schemes sponsored by the Waterways Board as well as by some local authorities were used to provide an additional labour force and, despite initial difficulties, the schemes were widely regarded as a success. Wiltshire and Berkshire County Councils also set up technical working parties that included representatives from the Kennet and Avon Canal Trust, the British Waterways Board, and all the district councils through which the Kennet and Avon passed. The conclusions reached by these groups was that work should continue until full restoration was achieved and the various councils should encourage and assist. Wiltshire County Council agreed to provide an adequate water supply to the canal summit, and Avon, Wiltshire and Berkshire County Councils set up officer advisory groups to provide civil

136 (*Top Right*) *The* Charlotte Dundas *was an early trip boat built from an old dredger pontoon and fitted with a 30hp diesel engine and hydraulically powered paddle wheels, the only drives that could cope with the extensive surface weed encountered in the early days of restoration, c.1975.*

137 (*Middle Right*) *Avoncliffe aqueduct before restoration commenced.*

138 (*Right*) *Restoration begins at Dundas wharf, with the more modern crane towering over the iron wharf-side one.*

139 (*Above*) *An aerial view of pre-restoration Caen Hill showing the side ponds covered in weed. Also shown is the brick works, now completely demolished, where many of the bricks used in building the canal section were manufactured.*

140 (*Left*) *A photograph of pre-restoration Caen Hill showing the total dereliction of the locks.*

141 Devizes wharf before any restoration work had commenced. The long, low building in the distance would become the Canal Trust headquarters.

were under Canal Trust management, the Avoncliff aqueduct was repaired, and the Limpley Stoke dry section relined. A new lift bridge had been installed at Aldermaston, and passenger boats were in operation at Reading, Hungerford, Crofton, Devizes, Bradford-on-Avon and Bath. In addition, some commercial operations had started at Newbury, Kintbury, Bath and Bristol.

This was a phenomenal achievement and one of which the Canal Trust and its volunteers could be truly proud. The partly restored waterway was becoming quite a showcase exhibit and attracting many visitors, including dignitaries, officials and other members of tours organised by the governments of Holland and France. The Trust opened an interpretation centre at Devizes wharf, where the Navigation's history was displayed using photographs, documents

engineering and other advice and to assist the restoration process generally.

Clew notes that by the beginning of 1981 all locks between Bath and the Devizes bottom lock had been restored and a start had been made on Caen Hill flight. In August of that year, however, the Canal Trust decided that a guaranteed income of around £75,000 a year would be necessary if their ongoing restoration plans were to be achieved, and work would need to be concentrated on those areas that required the most restoration. Water supply problems were partly overcome by the installation of electrical pumps at Claverton.

In 1983 the Kennet and Avon Canal Trust, with its proud record of achievement, celebrated its 21st birthday. Clew notes that by this time some 44 locks had been repaired along the length of the waterway and two eliminated, whilst an additional 28 lock chambers had been restored at Devizes and some new side ponds excavated as well. The pumping stations at Crofton and Claverton were restored and operating again, wharves at Devizes and Bradford-on-Avon

142 A section of the canal during restoration showing steel pilings being inserted.

and information panels. At the same time a Trust archive was established and a Trust shop opened at the wharf. Later, the interpretation centre and archive were further developed to become a registered museum.

In 21 years of restoration the Canal Trust had raised well over £1 million, including nearly £174,000 given by local authorities. During that period the British Waterways Board had also spent £220,000 on repairs and maintenance, whilst the financial cost of volunteer effort was unknown. The restoration task was not yet complete, however, and significant levels of work remained. Money-raising events were still being organised by trust members and local authorities and other benefactors, including the Inland Waterways Association, continued to provide an additional flow of funds. With the assistance of workers occasionally provided by the Manpower Services Commission to

143 *Members of the Youth Division working on the canal bed in very wet and muddy conditions, using similar tools to those that the original navvies would have used.*

augment the efforts of volunteers and British Waterways staff, restoration and maintenance work continued steadily.

An anticipated date of 1988-9 was scheduled for the completion of restoration work and it was also felt that funds obtained or promised would be sufficient to cover the estimated £750,000 required. But numerous problems and limiting factors still affected progress. Lindley Jones recounts that in the spring of 1986 the British Waterways Board announced that the expected completion date could only be met if lock gates could be provided from a source other than its own workshops. The production facilities there did not have the capacity to manufacture the required 35 sets of bottom gates and six sets of top gates in time. The National Association for the Care and Rehabilitation of Offenders therefore agreed to sponsor a Manpower Services Commission-funded gate-making workshop at Shrivenham to produce some of the gates, whilst the Waterways Board agreed to manufacture the remainder.

In a fashion reminiscent of that employed by the original canal proprietors almost two centuries earlier, some sections of the re-emerging Kennet and Avon were opened as they were completed. By 1988, however, it was evident that funds over the estimated target of £750,000 would be necessary, and fundraising activity was again initiated. Towards the end of that year the British Waterways Board announced a major re-organisation to establish its administration on a regional basis and allow issues to be managed locally. It also changed its name to British Waterways.

Lindley Jones notes that, as restoration proceeded, the lack of a constant and reliable water supply to the canal became an increasing issue. An investigation by British Waterways resulted in a 1989 report that indicated which water sources could be used to overcome the problem, and showed back-pumping installations which would be required for this

144 *Volunteers working near a bridge and using lengths of timber to form walkways and barrow ways over the mud in the navvies' tradition.*

145 *Volunteer wheeling a load of 'muck' up the barrow way.*

to happen. The Canal Trust did not agree with the report's estimates of demand, however, and set up its own sub-committee to monitor and survey annual water usage. At the Trust's Annual General Meeting that year, British Waterways representatives reported that schemes required to overcome water supply problems would cost around £2.5 million, some of which would have to be raised by the Trust, and until the necessary work had been completed the use of locks by canal users would need to be restricted. It was also emphasised by the Chairman that, although restoration was on schedule for 1990, water supply problems meant the canal would not be a perfect leisure amenity, upon which every boater would immediately be able to move as and when they wished.

An increasing number of pleasure boats were using parts of the waterway, although the water problems resulted in lock transits taking longer than they would otherwise have. Restoration in other areas was progressing well, although unresolved problems with maintenance agreements and responsibilities involving British Waterways and local authorities along the canal's length were still causing concern.

To the great delight of all concerned, the start of 1990 heralded an announcement that Her Majesty The Queen had agreed officially to open the restored waterway in the summer. Lindley Jones remarks that this important deadline tended to concentrate the collective minds of both British Waterways and Canal Trust managers, who very quickly agreed a timetable for completing work in progress. They also started to plan the many arrangements required for the royal visit, and organise a rolling programme of associated celebrations.

The highly successful royal visit involved the Queen and her party meeting the Canal Trust's President and Chairman, together with the Chairman and Chief Executive of British Waterways. Queen Elizabeth then enjoyed a brief walk before meeting Canal Trust staff and volunteers and touring the museum. The opening ceremony was carried out at Caen Hill, where the Queen cut a tape that had been stretched across the canal. A plaque was unveiled and a commemorative bowl presented to the Queen who, after further introductions, was driven away in a glass-topped Rolls Royce.

The waterway was now navigable from one end to the other, although parts of it were still in a perilous condition. Sections of the canal bed, especially at its western end, were known to be unstable. As these sections had been classified as remainder lengths, existing statutes banned British Waterways from spending their funds on repairing them. Structural breaches here could be catastrophic. In addition, certain sections of early restoration carried out with an eye to economy now required extensive re-working. Continuing water supply problems at Bath, Wootten Rivers and the Caen Hill flight of locks, together with a general lack of facilities for boaters in many areas, also caused concern.

The size of the task was still enormous, but a decision to start by dealing with water supply issues at Caen Hill was quickly taken. The planned scheme for completing this initial work would cost around £1 million, a sum that British Waterways announced they could not provide. It would have to be raised by the Canal

Trust in conjunction with any local authorities that might be prepared to help. After much constructive debate, the eight riparian local authorities, together with the Canal Trust and British Waterways, formed the Kennet and Avon Partnership and between them raised the necessary funding. A back pumping scheme that allowed the locks at Caen Hill to open to boats whenever required was formally commissioned in the summer of 1996.

The next step was to install back pumping systems at Bath and Wootton Rivers, though securing funding for this and the other extensive work still required would be immensely difficult. However, the mutual trust, joint experience, and excellent working relationships that had developed within the partnership encouraged and enabled British Waterways and the Canal Trust, to make a submission to the newly formed Heritage Lottery Fund for funds to complete the restoration and secure the waterway's future.

146   *Queen Elizabeth aboard the* Rose of Hungerford *during the royal visit.*

A discussion of the survey and planning work, and of the comprehensive preparation that was necessary before the bid could be submitted, is outside the scope of this book, and readers interested in the matter should refer to Peter Lindley Jones' excellent book, which also describes the restoration process in much greater detail. In the autumn of 1996 the Heritage Lottery Fund announced that it was awarding a grant of over £21 million, to which more than £7 million match funding would have to be added. This would need to be jointly raised by members of the Kennet and Avon Partnership over a six-year period.

After preparing and agreeing programmes of ongoing restoration, and establishing management and supervisory arrangements, work on the final push was ready to commence. Lindley Jones notes that winter work contracts for 1997-8 were in place by the autumn of 1997 and successfully completed by the end of the period. Dredging, replacement and repair work were all carried out over the coming years, although mainly in the winter months so that the waterway could open to traffic each Easter and be fully available for use during the main tourist season. As work progressed, public meetings and exhibitions were arranged to explain details of the restoration and educate and inform members of the public.

During 2001, with completion very much in sight, it appeared that unforeseen structural problems and the requirements of new legislation to protect water vole habitats meant an overspend of around three per cent might have to be faced. The problem was eventually resolved after discussions with the funding body and an agreement that British Waterways and the Canal Trust would meet the likely deficit jointly. By 2002 restoration was complete, and in the spring of the following year the final completion was celebrated with a visit from HRH Prince Charles. At long last the Kennet and Avon was a living waterway once more.

147 *Clearance work continues, this time in a lock chamber.*

148 *Lock excavation using mechanical means. The lock cill and general construction can clearly be seen in this photograph.*

149   *Rebuilding a lock chamber and finishing the adjacent canal bed.*

The high quality of the planning, financial and constructional management, as well as the actual restoration work, meant that the restoration team was short-listed from among 200 projects for the British Construction Industry Civil Engineering Award and the Construction Best Practice Award. As Lindley Jones notes, although the honours were finally bestowed elsewhere, the inclusion in the short-list of five for such high profile awards was recognition enough of an important achievement. The team did however have the satisfaction of winning an award when the Engineering Council's Award for Engineers for 2001 was presented to them.

10 | A Splendid Amenity:
The Kennet and Avon Today

Like most navigable waterways, the Kennet and Avon links areas of high population, such as Reading, Newbury, Bath and Bristol, with smaller towns like Hungerford, Devizes, and Bradford-on-Avon. In addition, this most southerly of trunk waterways, incorporating a wide canal, passes through or close to many picturesque villages and hamlets and crosses some of the most idyllic and unspoilt areas of Britain.

Following the completion of restoration work in 2002, the Navigation was considered to be in better condition than it had ever been, including after its initial construction some two centuries before. Although certain stretches, such as that between Reading and Newbury, still provided challenges to the boatman, users could take advantage of numerous improvements. Gone were the back-breaking, manually operated swing bridges; these were replaced by more advanced structures boasting push-button controls and subsequent ease of movement. Care had also been taken during restoration to ensure that travelling on the Navigation could be rewarding, especially where nature was concerned, and, wherever possible, existing wildlife habitats had been protected and enhanced and new ones established. Native plant life was encouraged to grow along the banks and in the surrounding waterway margins.

Formed in the days of restoration, the public corporation known as British Waterways was,

and still is, tasked with caring for some 2,000 miles of waterway within the United Kingdom, including managing and maintaining the 87 miles of canal and rivers that constitute the Kennet and Avon. This public body is governed by many of the original Enabling Acts, as they relate to particular canals, together with a number of more recent Acts of Parliament, and sees its role as ensuring a sustainable and integrated network of waterways throughout Britain. As a consequence, it aims to provide maximum benefit and enjoyment to those who use the waterways.

Where the Kennet and Avon is concerned, British Waterways is not always able to fully maintain sections of the waterway that do not have cruiseway status and such problems are sometimes alleviated through assistance from, for example, riparian local authorities, and by help from other Government organisations. Additionally, the Kennet and Avon Canal Trust continues to raise money for specific projects and support British Waterways in fulfilling its objectives whenever suitable opportunities arise.

The Canal Trust's original aim was to restore the waterway, especially the canal section, to provide through navigation and a public amenity and this was finally achieved when the whole route was re-opened from Reading to Bristol. Since that time the Trust's role has moved away from restoration to focus on ensuring the Navigation is protected from neglect, abuse and inappropriate development. It also provides associated funding, either from its own resources or by raising funds and acquiring grant aid whenever possible. Of equal importance is the Trust's role in promoting the Kennet and Avon as a major national amenity that is freely available for the enjoyment of all. To assist with these processes it is organised into a number of branches, and regularly raises money by providing and operating various sites and activities along the Navigation, opening them to the public. These include the pumping stations at Claverton and Crofton, three trip boats, four wharf-side shops and tearooms, and a museum and archive at the Trust's headquarters in Devizes.

Although significant improvements have been made, the Navigation still appears very

much as it did in its commercial heyday, although now, of course, trading barges and working narrow boats have given way to pleasure craft. Despite this, the Navigation provides a form of living history for those interested in its heritage, and the visitor, in addition to viewing and examining aqueducts, bridges, locks and other related structures, can gain knowledge of the Navigation's history. One can view the numerous wharves and the pumping stations at Crofton, with its great steam-driven beam engines, and at Claverton, with its magnificent water-driven machinery. During the season both these unique installations can be seen in operation, worked by Trust members who, in the winter months, also arrange and carry out maintenance of the steam- and water-driven machinery.

The Trust museum, situated in what was originally a canal-side warehouse on Devizes wharf, provides more general information about the Navigation's history. Artefacts, exhibits and information panels provide visitors to the museum with an appreciation of the building, operation, decline and eventual restoration of the Kennet and Avon. A large photographic and document archive is also located in Devizes, where researchers are able to delve more deeply into the waterway's history. Additionally, a smaller collection of canal-related artefacts and information may be seen at Newbury wharf and at Blake's lock, situated to the east of Newbury, and a number of individual wharves also have information boards located

150   *Enhanced wildlife habitat on the canal section.*

151   *The Kennet and Avon Canal Trust headquarters and museum at Devizes wharf.*

152   *Some of the displays inside the Canal Trust waterways museum.*

153　*Pleasure boats passing through the locks at Caen Hill flight.*

on them. Provided by British Waterways in conjunction with the Canal Trust, these boards indicate by the use of illustrations and text the sort of activity that occurred at each individual wharf in its working heyday.

Commercial trading craft have disappeared, but pleasure craft of varying size and shape are very much in evidence. There has been a steady increase in such activity over the years, with marinas, chandlers and other associated organisations being established to support the growing number of narrow boats, cruisers and smaller craft that use the Kennet and Avon. Temporary and permanent moorings

are available from British Waterways, who also issue boat licences, and moorings are also available from private operators at marinas. Moorings have to be paid for, but allow local owners to keep their boats in relatively convenient and safe locations and provide facilities for visiting boats to stop briefly whilst undergoing cruises of longer duration.

For those interested in longer distance cruising, a junction on the Kennet and Avon Navigation at Reading allows boats to travel on to the Thames, then up the Oxford Canal to join the Midlands canal network. From here, boaters can venture even further north, to, for example, the Leeds and Liverpool and associated canals, or north-west on the Llangollen and Montgomery canals. It is also possible to join the canal networks of central and northern England and Wales by taking a westerly route down the Kennet and Avon, then along the tidal stretches of the River Avon to the Severn Estuary, before travelling up the River Severn and on to the Gloucester and Sharpness canal to Gloucester and beyond. Such a journey would be more difficult than the more easterly route and would require careful passage planning. Assistance from a river pilot might also be necessary to travel on the tidal and difficult waters of the River Severn and its estuary, where tides, adverse currents, high wind and choppy waters could create problems. This would be especially true for narrow-beamed, flat-bottomed craft with their low freeboard, poorer stability in waves and, in the absence of a keel, propensity to drift sideways in conditions where wind and tide are unfavourable.

Water festivals and events are periodically held at locations along the Navigation. Organisers encourage general visitors who may not be boaters to come along and enjoy these events, but water festivals in particular tend to revolve around narrow boats, cruisers and other similar craft. One regular event is different: the annual 125-mile Devizes to Westminster canoe race. This popular international competition started in 1948 and is of major importance in the canoeist's calendar, attracting upwards of 200 participants, as well as numerous supporters and other interested parties. Apart from two years when adverse weather conditions followed by a Foot and Mouth outbreak caused its cancellation, the race has taken place every year since. To participate successfully, canoeists not only need sufficient skill to make a fast transit on the water but also require physical fitness and stamina. In addition to the many hours spent paddling, they have frequently to lift their canoes out of the water and carry, or 'portage', them around locks before continuing their journey.

The Kennet and Avon's towpath is level and wide for the most part and forms an ideal route for cyclists and walkers as it follows alongside the Navigation, passing through towns,

154   *The Waterway Festival at Newbury wharf.*

155 *Two contestants photographed on the canal during the Devizes to Westminster canoe race.*

156 *A National Cycle Network signpost alongside the Kennet and Avon.*

villages, rolling downland and countryside. Considered suitable for all ages and abilities of cyclist, much of the path is in fact part of Route 4 of the National Cycle Network that extends from St David's on the western tip of the Pembrokeshire coast. Leisure walking alongside the Kennet and Avon is, according to British Waterways' statistics, a continually increasing activity, especially on the canal section, and is undoubtedly the best way to observe the varied flora and fauna that abound both on and alongside the water and in the surrounding areas. Angling also remains a popular pastime, and the towpath can be used to fish for a number of coarse species ranging from potentially large pike and carp to relatively smaller specimens such as roach and rudd. Virtually the whole length of the waterway is leased to various angling associations and fishing clubs, and fishing competitions can on occasion take priority, where the best fishing spots, or 'pegs', are fought over.

Because of its nature and location, the Navigation also provides an important site for wildlife and its conservation, and a number of non-statutory nature reserves exist along the waterway's length. Due to the great bio-diversity they contain, several of these sites have been classified as areas of special scientific interest. In addition to the varied flora that grows in and alongside the Kennet and Avon, vertebrate and invertebrate animals and diverse insects flourish along the waterway and its margins. These include rabbits, stoats and weasels, as well as less common mammals such as water voles, the diminutive water shrew and, in the dusk, various species of bat. Frogs, toads, lizards and snakes are all in evidence, but usually remain well hidden.

Rarer species of bird such as reed buntings, barn owls and red kites, can occasionally be seen, whilst herons, moorhens, coots and a host of other waterfowl are more common. Kingfishers are not uncommon, although their small size, shy nature, and speed when flying ensures that sightings are infrequent, usually just a blur of colour seen in the corner of one's eye. Dragon and damsel flies can often be observed as they hawk across the water and reed beds, and the chalk downs through which the Kennet and Avon passes are host to numerous species of butterfly, many of which can be seen along the Navigation during the daytime, giving way to various species of moth as dusk descends.

The Kennet and Avon supports a great deal of tourist and leisure activity, and a report prepared for British Waterways in 2006 indicated significant increases in visitor numbers over previous years. The statistics also showed that whilst visits to hire and trip boats were on the increase, the limited number of moorings meant there had been little movement in the number of privately moored boats.

The increase in visitor numbers was thought to be due to informal visits to the waterway's towpath, which had increased at a greater rate than countryside visits nationally. It is likely these upward trends will continue and, as plans to increase the number of marinas and mooring facilities are in the pipeline, the pressure on moorings should in time be reduced.

Apart from certain small wharf-side developments in some of the towns through which the Kennet and Avon passes, larger projects that could radically and perhaps adversely affect the Navigation seem to be unlikely, especially as for most of its length the waterway passes through rural areas and green field sites. A small number of major schemes associated with large conurbations, which will to some extent involve the Kennet and Avon, have been agreed, and in some cases initial design and other work has already started.

The Western Riverside Scheme at Bath is centred on a largely derelict industrial site

157 *Fishermen at the top of Caen Hill. Walkers and a cyclist are also using the towpath, despite the rain.*

158   *A kingfisher in position to dart into the canal for its next meal.*

alongside the River Avon and is planned to provide extensive residential accommodation, with the possibility of riverside walks and river transport with ferries, water taxis and associated facilities being incorporated within the overall plan. The second of the major undertakings is part of West Berkshire Council's vision for Newbury town centre. Whilst incorporating housing, the plans are more focused on leisure, and have been the result of a partnership between district and town councils, together with local businesses and leisure and voluntary groups. They recognise that although the Kennet and Avon Navigation is one of Newbury's main attributes, it is not well utilised or presented within the town and that in the future this important asset should feature to a much greater extent. The Newbury development is planned in two phases. The first, scheduled for completion in 2010, will, in addition to providing waterside apartments and a park pavilion, incorporate a new water-based leisure facility. Plans for the second phase include two inter-connected water basins, one of which would be used for overnight moorings, together with a restaurant, inn and space that could become a possible waterways museum. It is envisaged this phase will be completed by 2012.

Businesses and waterway-associated ventures along the length of the Navigation significantly benefit from the tourism and leisure activities that it attracts. If restoration had not taken place it is unlikely that the waterway would be the significant facility it is today, and its wealth-creating potential would, in all likelihood, have been much reduced as a consequence. Instead, the restored Navigation continues to deliver benefits to the local economies through which it passes, and it is interesting to note that in real terms these are likely to outweigh those that were in place when the Kennet and Avon was a successful trading concern.

BIBLIOGRAPHY

PRIMARY SOURCES

Unless indicated otherwise, all primary sources referred to are held at the Kennet and Avon Canal Trust
Archive at Devizes.

By-Laws of 1827 for the Kennet and Avon Canal Navigation (1827)

Copy of an *Act to make the River Kennet in the County of Berkshire navigable from Reading to Newbury*,
London (1715)

Copy of an *Act for making the River Avon, in the Counties of Somerset and Gloucester, navigable from the City of
Bath to or near Hanham Mills*, London (1712)

Copy of an *Act for making a navigable canal from the River Kennet at or near the town of Newbury, in the County
of Berkshire, to the River Avon, at or near the City of Bath; and also certain navigable cuts therein described*,
London (1794) and on the River Avon, Bristol (1785)

Kennet and Avon Canal Committee Minute Books, Marlborough, various dates

Kennet and Avon Canal Committee, *Management Reports to Proprietors*, Marlborough, various dates

Priestley, J., *Historical Account of the Navigable Rivers, Canals, and Railways Throughout Great Britain*, London
(1831)

Rennie, J., *Reports to the K&A Canal Company Committee*, various dates

Robbins, Lane, and Pinniger, *The Honeystreet Wharf Accident and Sick Fund*, Honeystreet (1896)

Shaw, J.M., Haven Master, *Letter concerning rock problems at Hungroad*

HMSO, *British Waterways – Recreation and Amenity*, London (1967)

HMSO, *The Facts About the Waterways*, London (1965)

HMSO, *The Future of the Waterways*, London (1964)

HMSO, *Transport Policy White Paper*, London (1964)

Waylan, James, *Chronicles of the Devizes*, London (1839)

SECONDARY SOURCES AND FURTHER READING
BOOKS

Carr, E.G.G., *Sailing Barges*, London (1931)

Clew, K.R., *The Kennet and Avon Canal*, London (1969)

Elton, A., *The Pre-History of Railways*, Devon (1963)

Gaggs, J., *Book of Locks*, Princes Risborough (1975)

Gladwin, D.D., *The Canals of Great Britain*, London (1973)

Green, C., *Severn Traders*, Lydney (1999)

Hackford, C. and H., *The Kennet and Avon Canal*, Stroud (2001)

Hadfield, C., *British Canals, an Illustrated History*, Newton Abbot (1959)

Hadfield, C., *The Canals of South and South East England*, Newton Abbot (1969)

Hadfield, C., *The Canal Age*, Newton Abbot (1968)

Jackman, W.T., *Transportation in Modern England*, Cambridge (1916)

McKee, E., *Working Boats of Britain – Their Shape and Purpose*, London (1983)

Lindley-Jones, P., *Restoring the Kennet and Avon Canal*, Stroud (2002)

Newton, J., *The Falls of Burbage Wharf*, Huddersfield (1995)

Paget-Tomlinson, E., *The Illustrated History of Canal and River Navigations*, Ashbourne (2006)

Paget-Tomlinson, E., *Waterways in the Making*, Leominster (1996)

Pellow, T., *Waterways at Work*, Hereford (2000)

Rolt, L.T.C., *Navigable Waterways*, London (1969)

Smith, D.J., *Discovering Craft of the Inland Waterways*, Aylesbury (1977)

Sullivan, D., *Navvyman*, London (1983)

Willan, T.S., *River Navigation in England*, London (1936)

Wilson, D.G., *The Making of the Middle Thames*, Bourne End (1977)

JOURNALS, ARTICLES, REPORTS

Berry, R.W., 'Newbury Barges' in *The Butty, Journal of the K&A Canal Trust*, No. 171, Devizes (2005)

Berry, R.W., 'The Kennet Barge' in *The Butty, Journal of the K&A Canal Trust*, No. 172, Devizes (2005)

Berry, R.W., 'The Lost Wharves of the Kennet and Avon' in *The Butty, Journal of the K&A Canal Trust*, No. 174, Devizes (2006)

Berry, R.W., 'Bath's Lost Wharves' in *The Butty, Journal of the K&A Canal Trust*, No. 177, Devizes (2006)

British Waterways, 'Our Plan for the Future 2002 to 2006', Watford (2002)

British Waterways, 'Economic Evaluation of the Restoration of the Kennet and Avon Canal', Birmingham (2006)

Bryant, P.D., 'Did the Railways Kill the Canals – A Case Study of the Kennet and Avon Canal and the Great Western Railway', unpublished MA dissertation, University of Reading (1999)

Buchanan, B., 'The Avon Navigation and the Inland Port of Bath' in *Bath History*, Vol. VI, Bath (1996)

Dalby, J., Various articles in *The Butty, Journal of K & A Canal Trust*, Devizes

ECOTEC, 'Economic Evaluation of the Restoration of the Kennet and Avon Canal', Birmingham (2006)

Hackford, C., 'G.W.R. – Saviour or Destroyer' in *The Butty, Journal of the K&A Canal Trust*, No. 146, Devizes (1996)

Hackford, C., 'People of the Kennet and Avon Canal', *Berkshire Family Historian* (March 2002)

Harris, D., 'Reading's Lost Wharves' in *The Butty, Journal of the K&A Canal Trust*, No. 178, Devizes (2007)

Harris, D. and Naylor, B., 'First to Westminster' in *The Butty, Journal of the K&A Canal Trust*, No. 178, Devizes (2007)

Kennet and Avon Canal Trust, 'Aims and Objectives', Devizes (2006)

Saady, D., 'A Vision for Newbury Wharf' in *The Butty, Journal of the K&A Canal Trust*, No 178, Devizes (2007)

Skempton, A.W., 'The Engineers of the River Navigations 1620-1760', Transactions of the Newcomen Society (1953)

Smith, C., 'Building Stone from Combe Down and Odd Down', Bristol

Willan, T.S., 'Bath and the Navigation of the Avon' in *Somersetshire Archaeological Society Proceedings*, Bath (1936)

INDEX

Numbers in **bold** refer to illustrations